Monographs on Astronomical Subjects: 5

General Editor, A. J. Meadows, D.Phil.,

Professor of Astronomy, University of Leicester

The Nature and Origin of Meteorites

In the same series published by Oxford University Press, New York

The Nature and Origin of Meteorites

D. W. Sears

Department of Physics
University of Birmingham

Monographs on Astronomical Subjects: 5

OXFORD UNIVERSITY PRESS
NEW YORK

Library of Congress Cataloging in Publication Data

ᴵⁱ Sears, D W
    The nature and origin of meteorites.
    (Monographs on astronomical subjects; 5)
    Bibliography: p.
    1. Meteorites. I. Title.
QB755.S4 1978   523.5   78-10127
ISBN 0-19-520121-3

First published in the United Kingdom 1978 by Adam Hilger Ltd, Bristol
Adam Hilger is now owned by The Institute of Physics.
First published in the United States 1978 by Oxford University Press,
New York.
Printed in Great Britain by J. W. Arrowsmith Ltd, Bristol BS3 2NT.

Preface

Meteorites occupy a unique place in the study of man and his environment, bearing as they do on such diverse questions as the formation of the solar system and the origin of life on Earth. The study of meteorites therefore involves every scientific discipline. I have tried to reflect this by breaking the subject down into a few basic chapters in which the contributions of the geologist, chemist, physicist and astronomer are concentrated, but there is of course considerable overlap. It is hoped that this monograph will therefore complement some already established books which deal with the subject matter in greater depth, but from just one point of view.

Most of this monograph was written while I was a member of the Metallurgy Department of the University of Manchester. My first debt of gratitude is therefore to that institution and the many persons there who helped me in various ways. I am also pleased to thank Professor A J Meadows of the University of Leicester for his guidance over the work as a whole, Dr R Hutchison of the British Museum (Natural History) and Dr E Scott of the University of Cambridge for reading and improving much of the text, and Dr R S Clarke of the Smithsonian Institution for providing many illustrations and helping with their selection. Many other persons have helped with information and illustrations, and I extend to them all my warmest thanks. Finally, I feel bound to acknowledge the contribution of an eternally optimistic wife.

D. W. Sears

University of Birmingham

Contents

1. Historical Introduction

1.1. Thunderstones and the Origins of Modern Meteorite Research

The name of the town is Hatford, some eight miles from Oxford. Ouer this Towne, vpon Wensday being the ninth of this instant Moneth of April 1628, about five of the clocke in the afternoone this miraculous, prodigious, and fearfull handyworke of God was presented. It beganne thus: First, for an onset, went off one great cannon as it were of thunder alone, like a warning peece to the rest that were to follow. Then a little while after was heard a second; and so by degrees a third, vntill the number of 20 were discharged (or thereabouts) in very good order, though in very great terror.

... at the end of the report of euery cracke, or cannon-thundering, a hizzing Noyse made way through the Ayre, not unlike the flying of Bullets from the mouthes of great ordnance; and by the judgement of all the terror-stricken witnesses they were Thunderbolts. For one of them was seen by many people to fall at a place called Bawlkin Greene, being a mile and a half from Hatford.

The form of the Stone is three-square, and picked in the end: In colour outwardly blackish, somewhat like Iron: crusted over with that blacknesse about the thicknesse of a shilling. Within it is a soft, of a colour, mixed with some kind of minerall, shining like small peeces of glass.

What the good people of Hatford had observed, on 16 April 1628, was the fall of a chunk of rock from space. Small grains are being caught by the Earth all the time. They give rise to 'meteors', or 'shooting stars', as they burn their way through the atmosphere. Only occasionally does a large body fall, and then a recoverable object, a meteorite, may survive. There can be few ways in which we witness cosmic phenomena more dramatically than during the fall of a meteorite to Earth. The meteorite which fell at Hatford was unfortunately not preserved (Lockyer 1890). The oldest meteorite seen to fall, and which is still in our collections, fell on the village of

1

Ensisheim in 1492; and for most of its terrestrial sojourn has hung in the town hall there. Its fall was depicted in a contemporary engraving (see Heide 1964). A fireball was observed over Battenheim, and after a violent explosion and much commotion the stone fell on Ensisheim. The populace were excited and terrified; the learned were mystified. Under instructions from Emperor Maximillian, it was hung in the village church and was taken to be an omen of ill for the Emperor's Turkish opponents.

We know that meteorites have been seen to fall and have been collected since ancient times: the oldest record comes from the Chinese. The Egyptians collected them, and they have been found in pyramids: indeed, the hieroglyphic symbol for meteoritic iron found in tombs means literally 'heavenly iron'. Both Greek and Roman authors have described the fall of meteorites and their subsequent treatment. Many meteorites were worshipped, and one was given a royal procession into Rome.

The modern science of meteoritics began in the eighteenth century. In 1768, the Abbé Bachelay was presented with a stone which, it was claimed, had fallen from the sky: this time at the town of Lucé in France. The Abbé collected the comments of witnesses and wrote a description of the stone, which he passed onto the French Academy of Sciences. The Academy set up a three-man commission of enquiry, consisting of Lavoisier, Cadet and Fougeroux, and they performed the first chemical analysis on a meteorite. They recognised pyrite (iron sulphide) to be present, and concluded that the stone was terrestrial pyrite. The phenomena of fall and the black fused crust of the stone, they said, were caused by its being struck by lightning (Fougeroux et al 1772). Their report was generally accepted, and their opinion widely adopted: consequently the stones became known as 'thunderstones'. Such stones were therefore considered nothing out of the ordinary, and several people advocated the destruction of meteorite specimens. The scientific fraternity are said to have 'made merry over the credulity' of people who imagined the stones to have fallen from the heavens' (Zittel 1901).

The middle and late eighteenth century saw the recovery of a number of large masses of native iron. One of the earliest was reported by Peter Simon Pallas, a German natural historian who had been invited by Catherine II to lead an expedition to the vast uncharted regions of Russia. Pallas undertook two lengthy expeditions, each of which culminated in massive volumes of writings and collections of specimens. The first began in the spring of 1768 and its purpose was to explore the empire east of the Volga. Pallas spent the winter of 1771–2 at Krasnojarsk, where he heard of a large strange mass of native iron which had been found by a blacksmith in 1749. This man had moved it to his place of work from the Yenisei River, some 32 km away. Even at this time there were reports that described the mass as having fallen from the sky. Pallas (1776) wrote a lengthy description of the iron and sent this and a piece of it, which he had removed with great difficulty, to the Academy of Sciences at St Petersburg. Five years later, soldiers were sent to collect the main mass of the iron (figure 1.1). As a consequence of the size of the iron, and the general interest in Pallas' travels, samples became widespread: at this time, all gentlemen of learning possessed mineral cabinets, and none was complete without a sample of the Pallas iron.

Several similar masses of native iron were found all over the world, notably in Argentina and South Africa. Perhaps because of the interest in these native irons, and the obscure legends that they had fallen from the sky, there was a considerable increase in the number of reports of meteorite falls at the end of the eighteenth century. Beginning in 1789 with Barbotan and ending in 1803 with L'Aigle, no less than nine meteorites fell, all of which became well known in Europe, and were well observed, documented and discussed. In 1794 the fall of the Siena meteorite caused a stir among the English community at Naples. The reaction in London became even more marked when in the following year, a meteorite which had just fallen on Wold Cottage, Yorkshire, was put on show in the 'Gloucester' coffee house. In 1796, a fall occurred in Portugal; and in 1798 falls occurred at Salles in France, and Benares in India. From this time, it became the convention to name a meteorite after its place of fall or where it had been found.

Throughout this barrage of meteorite falls a number of

Figure 1.1. The Pallas Iron, a large mass of iron found by P S Pallas on one of his scientific expeditions to Siberia. It soon became an object of much interest and may have indirectly contributed to the beginning of the scientific study of meteorites. It is now known to be a stony-iron meteorite ('pallasite'). Engraving from Pallas' *Reisen durch verschiedene Provinzes des Russischen* (1776).

books, articles and pamphlets had inevitably been published discussing their origin. For example, Edward King, who was trained in law and had written on an enormous range of subjects, produced one of the best known works on this topic. It was a pamphlet entitled *Remarks concerning stones said to have fallen from the clouds . . .* in which King advocated the idea that the stones were concretions of volcanic dust in the atmosphere. The most authoritative work, however, was written by E F F Chladni, an established physicist and the discoverer of 'Chladni's figures', patterns produced in sand on a vibrating plate. Chlandni's *Über den Ursprung der von Pallas gerfundenen Eisenmasse* was published in 1794: in this book he described the Pallas iron and four other native irons, and he collected together many reports of stones which had fallen from the sky. The irons in no way resembled terrestrial rocks, and had obviously been heated by natural means. Chladni therefore linked the numerous stories concerning falling stones with these unusual masses of native iron, and suggested that both had an extraterrestrial origin. Apart from the very unreliable and poorly documented legends about the Pallas iron, Chladni lacked any real link between the irons and the stones, since it was the stones which fell from the sky, but the irons that were unusual.

It was inevitable that, with scientific interest raised to such a pitch, chemical analyses would be undertaken. Sir Joseph Banks, as President of the Royal Society, had acquired several meteorites, and was convinced by their similarity in appearance that, wherever they fell, they were of common origin. He persuaded an English chemist, Edward Howard, to carry out chemical analyses. Howard (1802) worked with a French mineralogist named Bournon and soon found that it would be necessary to examine separately the various phases present in the meteorites. Unlike terrestrial rocks, stony meteorites contain millimetre-sized fragments of metal (which on Earth would rust away very quickly), curious globules of stony material and a thin black skin (the fusion crust). We now know that the black skin is produced by the heat produced on entry into the Earth's atmosphere. In the metal, Howard found a considerable quantity of nickel, which is rare on Earth, and had only been discovered in 1751: its chemistry was still novel. The presence of nickel linked the stones with native irons, in which the French chemist Proust had also found nickel a year or so previously. This was immediately confirmed by the eminent continental chemists, Vauquelin and Fourcroy in France and Klaproth in Germany. The stony meteorites were not only similar to each other in that their metal contained nickel, but also in the proportions of their other components, the globules and their crusts. The similarities between stones which had fallen all over the Earth finally convinced the scientific world that meteorites really had fallen from the sky and had a common origin (Sears 1975b). To underline the point, on 26 April 1803, a shower of over 3000 stony meteorites fell on the French town of L'Aigle, where it was witnessed by a member of the French Academy.

1.2. The Early Nineteenth Century

After hearing of the chemical analyses that had been carried out, LaPlace, Biot and Poisson suggested that the meteorites could have come from the Moon. This opinion remained the most popular until the middle of the nineteenth century. Its popularity probably stemmed from Herschel's alleged observation of active lunar volcanoes on the Moon in 1787. However, the idea that meteorites were terrestrial in origin did not die overnight: another major theory was that the meteorites were concretions in the atmosphere of matter which had been degassed from Earth. The proof of an extraterrestrial origin has been generally assumed to have come in 1833, when the radiant (the apparent point of origin in the sky) of the Leonid meteor shower was observed not to move with the Earth as it rotated on its axis. (At this time meteors and meteorites were thought to be the same phenomenon.)

The number of elements known to be present in meteorites rose steadily throughout the first few decades of the nineteenth century, as techniques developed and additional analysts entered the field. The early chemical work culminated in that of J J Berzelius, who analysed numerous meteorites of various types with a reliability and accuracy which was exceptional for the time. Berzelius (1834) established that although the majority of meteorites were remarkably similar, three (Stannern, Juvinas and Jonzac) showed widely differing properties. The mineralogist Gustav Rose (1825), a student of Berzelius, had attempted to identify the minerals in meteorites using the infant science of crystallography. He concentrated on the three meteorites which Berzelius later singled out as chemically different, because of their comparatively coarse texture, and found that they closely resembled terrestrial volcanic rocks. This observation was consistent with the lunar volcanic idea for the origin of meteorites, so that Berzelius was able to summarise the situation with the following words.

> We shall see hereafter that the greater number of meteoric stones resemble each other so much in their composition that they may be considered to come from the same mountain, that is, from the central culmination point of that side of the moon which is always turned towards the earth. A small number only present a different appearance, and it is therefore probable that these proceed from mountains situated on other parts of the moon.

The native irons also came under close scrutiny following the establishment of the authenticity of meteorite falls. In 1804 Alois von Widmanstätten in Vienna and William Thomson in Naples independently discovered that, when sectioned, polished, and etched with acid, many of the irons displayed a characteristic pattern (Smith 1962). Schreibers produced a set of prints illustrating meteorites, some of which showed the 'Widmanstätten pattern' (figure 1.2). The techniques for their

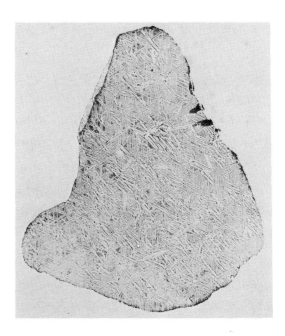

Figure 1.2. Direct typographical imprint of the Elbogen iron meteorite showing the Widmanstätten pattern (from Schreibers 1820). Elbogen is now recognised to be a member of class IID with a fine octahedrite structure.

reproduction were crude, but highly effective. Schreibers and Widmanstätten carefully etched the polished surface of the iron and used it directly as a printing block. Until the development of the microscope, little advance could be made from such studies. However, when Chladni (1819) drew up a catalogue of meteorites, he distinguished between those irons which showed the Widmanstätten pattern and those, like the Pallas iron, which consisted of a network of metal enclosing silicates.

1.3. The Mid-Nineteenth Century: Fall Phenomena, the Asteroidal Hypothesis and the Increased Pace of Research

Lightning, with the associated thunderclaps, could readily explain most of the phenomena associated with a meteorite fall, although when it was shown that meteorites were extraterrestrial, the light and sound phenomena were difficult to understand. It seemed most likely that something was burning. Aristotle, and much later Halley, had suggested that flammable vapours were the cause of meteors, and Brande suggested in 1817 that the upper atmosphere was somehow ignited by the passage of the meteorite. However, some early, and rather poor, determinations of the height of fireballs appeared to show that ignition occurred well beyond the atmosphere. In 1837 Poisson was therefore led to suggest that a fireball was an electrical discharge, triggered by the meteorite's passage. The development of thermodynamics and the work of Joule and others meant that much of the atmospheric behaviour of meteorites was understood by the middle of the century. A leading interpreter in this field was W K Haidinger, Director of the Hof Mineralien Cabinet in Vienna, who believed that a large volume of heated air was formed as the kinetic energy of the meteorite was dissipated (Haidinger 1869). When low in the atmosphere, the object was completely slowed down and fell only under gravity. The explosions which were frequently heard occurred at this height, and were due to air filling the void created behind the stone. The fusion crust was a natural result of superficial heating. These views were, in time, generally accepted: the only feature we would now dispute is the cause of the explosion (§ 2.2.1).

After the middle of the century, the consensus of opinion shifted from the lunar hypothesis to the idea that meteorites originated among the asteroids. This idea had been preferred by several investigators since earlier in the century; for example, by Chladni, Olbers and Brandes. After the first few decades of the nineteenth century, few astronomers accepted that lunar volcanoes were still active. Moreover, better determinations of meteorite entry velocities were frequently 'planetary'; that is, too high for a lunar origin. In addition, the number of known asteroids had grown to more than 30, and it was taken for granted that these represented a fragmented planet. Hence, it seemed inevitable that some pieces would find their way to Earth. The author of at least one major work (von Humboldt 1849) had no hesitation in calling meteorites 'the smallest of all asteroids'. Again there was no unanimity of opinion, however, and the Americans Shepard and Lawrence Smith were notable for their opposition to the idea of an asteroidal origin.

Perhaps the major feature of meteoritics around the middle of the nineteenth century was the marked increase in the number of studies and the number of observed falls reported (figure 1.3). A major factor in both cases was the growth

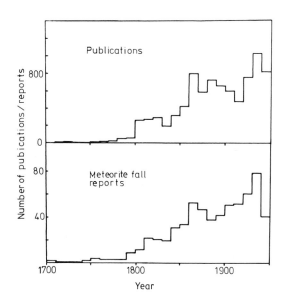

Figure 1.3. Histograms showing the growth in the number of publications concerning meteorites (data from Brown *et al* 1953) and the number of observed falls (data from Hey 1966).

in population and in the level of education; but the increase in the number of studies, which was more abrupt, was caused by a number of additional factors of which we may mention three: the growth of major collections, the establishment of professional research posts, and the introduction of thin sections. The collections in Vienna and Berlin had grown steadily since the beginning of the century, but the collections in Paris and London had a more chequered history. The rivalry between London and Vienna reached heated proportions, and lists were published comparing the two collections in terms of numbers, weight and quality. In 1862 the comparison stood at 120 stones and 68 irons in Vienna, and 111 stones and 79 irons in London.

Thin sections are slices of rock ground to about 0·03 mm in thickness, so that it is possible for light to pass through them and for their structure to be examined under the microscope. Some early observations are shown in figure 1.4. The 'curious globules' discovered by Bournon appear as circles with various textures, and some are most attractive. Rose (1863) gave the name 'chondrules' to these globules. Sorby (1864) recognised the textures as being typical of quenched systems and, since the chief characteristic is their rounded shape, suggested that they could be droplets of matter which had been splashed out from the Sun. Solar studies were most active at this time, and solar flares had been recognised. Sorby also concluded from a number of features—for example, cavities in the silicates filled with glass, and the overall texture—that the rocks were volcanic.

An innovation made soon after Sorby's work was the introduction of a set of polarising filters in the microscope. The crystalline nature of minerals causes them to polarise light, and the use of filters enables the detection of this phenomenon, as it gives rise to attractive diagnostic colorations and other effects. In this way, Maskelyne (1870) was able, for example, to identify many minerals present in meteorites, and to discover some unique to them, without the use of chemical analysis (which the fire regulations at the British Museum virtually prohibited).

1.4. The Late Nineteenth Century: Classification and Gas Analysis

It was the size of the collection available to him that enabled Rose, who was responsible for the Berlin collection, to achieve a high degree of success in the classification of meteorites in

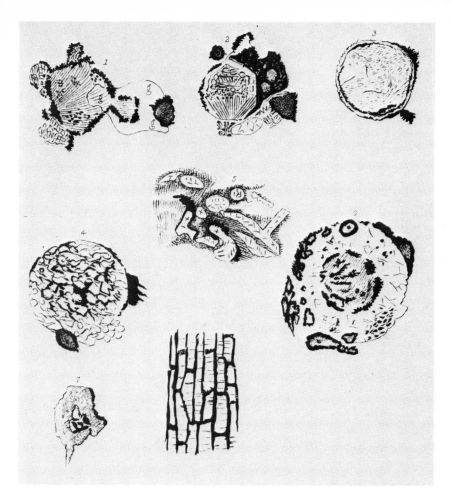

Figure 1.4. Early microscopic observations of the textures in thin sections of the Knyahynia stony meteorite. They are all sketches of chondrules, or details in chondrules, except at bottom left, where a metal grain enclosing silicates is shown (Kengott 1869).

1863. His was the third major attempt. The early work distinguished between irons and stones (actually the major aim was rather to establish a link between them). Chladni, in his 1819 catalogue, distinguished between spongy irons (such as the Pallas iron), and irons showing the Widmanstätten pattern. Shepard's was probably the earliest comprehensive scheme; it divided meteorites into two classes (metallic and stony), each of which was divided into three 'orders' and, finally, into a variable number of 'sections'. The result was 16 groups, membership being governed mainly by texture and analogy with terrestrial volcanic rocks. Riechenbach's scheme resulted in nine classes of families and groups based mainly on superficial features such as colour (for example, second-order, somewhat bluish stones). Rose's classification was based on mineralogy and composition and became the most popular (Rose 1863). After modification and enlargement by Tschermak (1883) and Brezina (1904), this system now forms the basis of modern classification. It is described in table 1.1.

Table 1.1. The Rose–Tschermak–Brezina classification scheme and its more recent equivalents[1].

Rose–Tschermak–Brezina	Prior–Mason	Alternatives used here
	Stones: silicates prevailing[2]	
Achondrites[3]	Chladnite, Chl ⎰Diogenites[4]	
	Veined chladnite, ⎱Hypersthene	
	Chla achond.	
	Bustite, Bu ⎰Aubrites	
	⎱Enstatite achond.	
	Eukrite, Eu Eucrites[5]	
	Howardite, Ho ⎱ Howardites	
	Breccia-like ⎰	
	howardite, Hob ⎱	
	Amphoterites, see [6]	LL chond.
	Am	
Chondrites	Cho, Choa, Cw, ⎫	
	Cwa, Cbw, Ci, ⎮	
	Cia, Cib, Cg, ⎮ Ol-bronzite	H chond.
	Cga, Cgb, Co, ⎮ chond.	
	Ct, Cs, Csa, Cc, ⎬ Ol-hypersthene	L chond.
	Cca, Ccb, Cco, ⎮ chond.	
	Ccn, Cck, Ccka, ⎮	
	Cckb, Ck, Cka, ⎮	
	Ckb, U ⎭	
	⎰Ureilites[7]	
	⎱Ol-pigeonite	
	K, Kc, Kca ⎰Carbonaceous	CB chond. CV
	⎨ chond.	chond.
	⎩Pigeonite chond.	

12

Table 1.1.—*cont.*

Rose–Tschermak–Brezina		Prior–Mason	Alternatives used here
Enstatite–anorthite chondrites	Cek	Enstatite chond.	E chond.
Siderolites[8]	Mesosiderites	{ Mesosiderites Pyroxene–plagio- clase stony-irons	
Iron meteorites: metallic constituents prevailing or alone			
Lithosiderites[9]	Pk, Pr, Pi, Pa	Pallasites, Ol stony-irons	
Octahedrites	Off, Ofv, Of, Ofe, Om, Ome, Og, Oge, Ogg, Obn, Obk, Obc, Obz, Obzg	Off, Of, Om, Ogg, Og	IAB, IC, IIAB, IIC, IID, IIE, IIIAB, IIICD, IIIE, IIIF, IVA, IVB, anom. irons
Hexahedrites	H, Ha, Hb	Hexahedrites	
Ataxites	D, Dsh, Db, Dl, Dn, Dp, Dm	Ni-rich ataxites Ni-poor ataxites[10]	

[1] Chondrites and achondrites are abbreviated chond. and achond. Olivine is written as Ol.

[2] The Prior–Mason classification and current practice make a third division of stony-irons ('siderolites' plus 'lithosiderites').

[3] Additional minor classes: angrites (augite achond., Angra dos Reis); chassignite (Ol achond., Chassigny), rodites (now considered to be diogenites), shergottites (Shergotty and Zagami).

[4] Diogenites, aubrites, ureilites and the chassignite: these are collectively referred to as Ca-poor achondrites.

[5] Eucrites and howardites are collectively called pyroxene–plagioclase achondrites and, together with nakhlites and the angrite, are referred to as Ca-rich achondrites.

[6] Amphoterites are now recognised to be chondrites and are classified as LL chondrites.

[7] Ureilites are now considered to be achondrites.

[8] Additional minor classes: grahamites (now considered to be mesosiderites), and the lodranite (bronzite–olivine stony-iron, Lodran).

[9] Additional class: siderophyre (bronzite–tridymite stony-iron, Steinbach).

[10] Now recognised to be thermally altered hexahedrites, and the former Ni-rich ataxites are simply termed 'ataxites'.

We will not dwell at length on the tortuous and sometimes confusing route by which modern classification schemes have emerged, except to mention a few basic points. Meteorites are divided into stones, stony-irons and irons depending on their major constituents. Berzelius (1834) first classified the stones into two groups on compositional grounds. Rose (1863) observed that chondrules were abundant in one group, but not in the other, and named them 'chondrites' and 'achondrites', respectively†. Stony-irons also require subdivision. In fact, this subdivision is so marked that in the Rose–Tschermak–Brezina system some stony-irons are placed with the stones and some with the irons, although they both contain equal proportions of iron and stony material. Irons were subdivided into 'octahedrites', 'hexahedrites' and 'ataxites' on the basis of their structures as revealed in etched, polished sections.

All these subdivisions require further subdivision, although the need is less for the more recent schemes. Originally there were 76 classes, of which almost half had only one member. The system, especially for chondrites, revolved around the allocation of descriptive labels, and contained an enormous amount of information, but was large and unwieldy. Prior (1920) simplified the classification by relying entirely on composition, and this practice was followed by Mason (1962) who extended it to mineral chemistry. Despite this, both Mason and Prior used mineralogical names for their classes. Current practice, following Van Schmus and Wood (1967), is to use letters: E chondrites for 'enstatite chondrites', C for 'carbonaceous chondrites' and H, L and LL for the ordinary chondrites, depending on the amount of iron present (§ 3.2.1).

An important feature of late nineteenth century meteoritics was the development of gas analysis. Graham had discovered in 1867 that metal from the Leonarto iron evolved nearly three times its own volume of hydrogen gas when heated to dull redness for 2·5 h. With the discovery of hydrogen lines in stellar spectra by Huggins and Miller in mind, he breathlessly remarked that ' . . . the meteorite may be looked upon as holding imprisoned within it, and bearing to us, hydrogen of the stars'. Gas measurements were extended to other irons and stones, notably by the Americans Mallet and Wright, and carbon monoxide and methane were found to be additional gases in meteorites. This finding was thought to be

† In retrospect, naming these essentially chemical groups on textural criteria was unfortunate, because chondrites are now known to exist which do not contain chondrules, and at least one achondrite exists which contains chondrules.

particularly significant because of the discovery of carbon in comets. For a while it was customary, when there were data available, to quote the gases present in the report on a chemical analysis. However, in 1886 an elaborate set of results was published by the influential Sir James Dewar, who showed that the gases evolved were the product of a reaction between water, which had become absorbed by the meteorite, and the constituents of the meteorites. As a consequence, the interest in gases largely waned. One finding of major importance to come out of this work was the discovery by Ramsay in 1895 of the presence of argon and helium in iron meteorites.

1.5. The Early Twentieth Century

1.5.1. Meteor Crater and the Tunguska Event

In 1891 an American mineralogist and dealer by the name of Foote discovered that some fragments of iron he had acquired were meteoritic. He followed this up, and found that they had come from the sides of a massive crater, 1300 m in diameter and 175 m deep. The crater was situated in Arizona, and around it he found an additional 137 fragments. The crater (figure 1.5) drew immediate attention, but so also did the meteorite specimens, because in them, Foote found diamonds. This is still the subject of much interest, and will be discussed later.

G K Gilbert, Director of the United States Geological Survey, and his staff visited the site and, although Gilbert had advocated meteorite impact for the origin of the craters on the Moon, they decided that the Arizona crater was probably not meteoritic. They suggested that it was volcanic, or the result of collapse after subsurface water erosion, and that the meteorites were a coincidence. Despite this conclusion, which was undisputed by geologists, a mining engineer named D M Barringer remained convinced of the meteoritic origin of the crater. From 1902 to 1929 he tried to locate the main mass, which he believed to be buried under the crater. He was supported in this by the meteoriticists Merrill and Farrington but, in 1929, the year Barringer died, the astronomer Moulton demonstrated that no meteorite with sufficient energy to make

Figure 1.5. The Canyon Diablo meteorite crater in Arizona, which is 1300 m in diameter and 175 m deep. An idea of the scale can be obtained from the buildings at bottom right, and the roads which circle the crater and approach the rim. Notice that the interior walls of the crater are much steeper than those on the outside, and that the bottom of the crater lies below the level of the surrounding plains. An uneven layer of material, which was ejected when the crater was formed, covers the surrounding plains. (United States Information Agency.)

the crater could have survived the impact. Much controversy followed these calculations, but they established the meteoritic nature of the crater. A series of geophysical measurements between 1931 and 1951 also failed to locate the main mass; although, in addition to the enormous number of fragments found on the surface, the soil was rich in microscopic nickel-rich spherules (Krinov 1966).

The discovery of the massive crater in Arizona (known variously as 'Coon Mountain', 'Coon Butte', 'Meteor Crater', 'Canyon Diablo Crater' and the 'Barringer Crater') was not the only dramatic incident in meteoritics in the early twentieth century. In 1908 a massive explosion occurred in a sparsely populated region in Siberia near the Tunguska River. The force of the explosion was sufficient to produce seismic and barometric waves which were received all over Asia and even as far away as England. Prolonged twilights and reddened skies were seen over Europe and Asia, suggesting that a considerable amount of dust was present in the atmosphere. Because of the First World War and the Russian Revolution (and, perhaps, doubt about the reality of such a fantastic event), it was 1927 before an expedition led by L A Kulik reached the area of the impact. What they found is best described in Kulik's own words (Krinov 1966). While still 8 km from the point of impact he wrote:

> I cannot sort out my chaotic impressions of this excursion. Above all, I cannot really take in the whole majestic picture of this unique meteorite fall. A very hilly, almost mountainous, region stretches away tens of versts towards the northern horizon. From our observation point no sign of forest can be seen, for every thing has been devastated and burned, and around the edge of this dead area the young twenty-year-old forest growth has moved forward furiously, seeking sunshine and life. One has an uncanny feeling when one sees 20–30″ thick giant trees snapped across like twigs, and their tops hurled many metres away to the south.

Elsewhere in his diary Kulik wrote:

> ... it was very dangerous to walk through the old dead forest; twenty-year-old giants rotted at the roots were falling down on all sides. Sometimes they fell quite close to us, and we sighed with relief when we went down into the cauldron or a valley sheltered from the wind.

For the next four years, annual expeditions were made to the site, but no meteorite fragments or craters were found. In 1930 the astronomer Whipple suggested that a comet had struck the Earth at Tunguska and, although many other ideas have been proposed (some very exotic), the idea became and remains the most popular. As a consequence of these two highly dramatic events, meteorites were accorded a high degree of popular interest between the wars.

1.5.2. Analytical Spectroscopy

The development of spectroscopy as an analytical tool was a lengthy process. Qualitatively, it was in general use from the middle of the nineteenth century: it was spectrum analysis that permitted Bunsen's discovery of lithium in meteorites in 1861. Quantitatively, spectroscopy only came into widespread use in the second and third decades of the twentieth century (see, for example, Noddack and Noddack 1930). Two concepts emerged from this early analytical work which are still important in meteorite research. The first was Goldschmidt's classification of the elements, and the second was Russell's observation of the chemical similarity between the Sun and meteorites.

V M Goldschmidt was a geochemist particularly interested in the internal structure of the Earth. By the end of the nineteenth century it was universally accepted among astronomers that the asteroids represented fragments of a disrupted planet. This being so, the meteorites gave a unique opportunity to examine the internal structure of a planet: the data obtained could be used in explaining the internal nature and formation of the Earth. Starting from the observation that meteorites consist of irons, stony-irons and stones, and that the dominant phases within these are metal, sulphide and silicates, Goldschmidt (1929) developed a model of the Earth whereby a metal core was surrounded by a sulphide shell, a silicate mantle, a thin crust and an atmosphere. Such a structure could be made, Goldschmidt envisaged, by melting material of stony meteorite composition. As a model for the formation of the Earth, these ideas lasted until a few decades ago. More significant from our point of view was his observation that the elements showed differing tendencies to concentrate in each phase, and could therefore be classified as *siderophile* ('iron loving'), *chalcophile* ('sulphide loving'), *lithophile* (concentrating in the silicates) and *atmophile* (gases). Goldschmidt's classification for each element, as determined by him from analysis of different phases in meteorites, is given in table 1.2.

The American astronomer H N Russell was not the first to compare the composition of meteorites with that of the Sun, as qualitative comparisons had been made many times. However, spectroscopic analysis of the Sun and meteorites had reached a sufficient state of refinement by the 1920s for quantitative comparisons to become feasible (table 1.3). The conclusion Russell drew was that meteorites resembled the Sun much

18

Table 1.2. Goldschmidt's classification of the elements (Goldschmidt 1929)†.

Siderophile	Chalcophile	Lithophile	Atmophile
Fe, Co, Ni, P, (As), C, Pt, Ir, Os, (Pd), Ru, Rh, Mo, (W)	((O)), S, Se, Te, Fe, (Ni), (Co), Mn, Cu, Zn, Cd, Pb, (Sn), Ge, (Mo), As, Sb, Bi, Ag, Au, Hg, Pd, (Ru), (Pt), Ca, In, Tl	O, (S), (P), (H), Si, Ti, Zr, Hf, Th, F, Cl, Br, I, B, Al, (Ga), Sc, Y, La, Ce, Pr, Nd, Sm, Eu, Gd, Tb, Ds, Ho, Er, Tu, Ad, Cp, Li, Na, K, Rb, Cs, Be, Mg, Ca, Sr, Ba, (Fe), V, Cr, ((Ni)), Nb, ((Co)), Ta, W, U, Sn, (C)	H, N, (Cl), I, He, Ne, Ar, Kr, Xe

† Secondary, minor and uncertain tendencies in parentheses.

Table 1.3. Abundances of the elements in the Sun, Earth and meteorites (Russel 1929)†.

	Sun	Earth	Meteorites		Sun	Earth	Meteorites
Na	8·6	8·7	7·8	V	6·7	6·9	—
Mg	9·2	8·6	9·1	Cr	7·4	7·1	7·5
Al	7·8	9·2	8·2	Mn	7·6	7·3	7·3
Si	8·8	9·7	9·3	Fe	9·0	9·0	9·4
K	8·4	8·7	7·2	Co	7·4	5·8	7·1
Ca	8·3	8·8	8·1	Ni	7·8	6·8	8·2
Sc	5·3	3·0	—	Cu	6·8	6·3	6·2
Ti	6·9	8·1	7·0	Zn	6·7	5·9	—

† Logarithmic scale in arbitrary units.

more closely than they did the Earth's crust. In particular, the latter appears to be enriched in Al, Si and Ti and depleted in Mg. The significance of this discovery and current views on the subject will be reviewed in § 4.1.

1.5.3. Prior's Rules

A major theme of meteorite research has been, and to some extent still is, the reduction–oxidation relationships within the chondritic meteorites. This is important because a single redox sequence would imply a common origin under increasingly oxidising or reducing conditions. These investigations require a large number of reliable analyses: it was a long time before these were available, because meteorites are notoriously difficult to analyse. In 1878 Nordenskoild collected analyses of nine meteorites and found that when oxygen, sulphur and phosphorus were ignored, the samples were chemically identical. On the inclusion of these elements, the only difference between the meteorites was the extent to which iron was in the oxidised state (that is, we have a continual reduction–oxidation sequence). Nordenskoild's observations did not attract the attention they deserved, but when Prior made the same observations in 1916 the importance of these ideas became generally recognised. Prior (1916) formalised the relationships in two rules:

> ... the less the amount of nickel–iron in chondritic stones, the richer it is in nickel and [secondly] the richer in iron are the magnesium silicates.

He believed it to be a continuous sequence, which he later extended to irons and stony-irons.

The strict validity of Prior's rules was disproved when a greater number of analyses became available. A considerable selection of data was necessary to reject poor analyses, the major problem being the effect of weathering which, being an oxidation process, would alone produce a relationship like Prior's rules. Out of 300 analyses, Urey and Craig (1953) found only 94 that were reliable. Within these, they observed that the chondrites fell into two groups which differed in total iron content: a high group (the H chondrites) and a low group (the L chondrites). To convert a member of one group into the other requires the addition, or removal, of iron, as well as oxidation or reduction. The extent to which the Prior-type rules apply within the groups is still open to debate (§ 3.2.2).

1.5.4. Early Inert Gas Measurements

The nineteenth century investigations using gas analysis proved, in the main, to be a major disappointment, but the discovery that there was an abundance of inert gases was of long-term significance. The major source of helium, which is very scarce in our own atmosphere, was thought from its terrestrial distribution to be the radioactive decay of uranium and thorium, and that its abundance must therefore be related to the age of the meteorite. Repeated measurements by Paneth *et al* (1928) showed that either meteorites were considerably older than the Sun, which was generally thought to be highly unlikely, or that they contained excess helium. Bauer (1947a, b, c) suggested that the excess helium was cosmogenic (that is, made by cosmic ray bombardment of the meteorites in space), in which case it should show a variation with depth. This was sought after, and was discovered by Paneth *et al* (1952). The physical processes to which meteorites have been subjected may therefore be considered twofold: radiogenic—related to the time the meteorite has existed as a single solid object (§ 5.2); and cosmogenic—related to its period of exposure to cosmic rays (§ 5.3). These two processes are not the only sources of inert gases in meteorites, because there also appears to be a certain amount of gas which is part of the original composition of the meteorite (the 'primordial' or 'trapped' component), and is

Table 1.4. Inert gases in Pesyanoe (10^{-8} cm^3 g^{-1} STP; Anders 1963).

	^{4}He	^{20}Ne	^{36}Ar
Pesyanoe	$7 \cdot 3 \times 10^5$	2150	159
Typical chondrites	1500	7	2
Atmosphere†	25	86	159

† Normalised to ^{36}Ar = Pesyanoe.

described in § 4.7. The third component was discovered by Gerling and Levski (1956), who found that the Pesyanoe meteorite was gas-rich (table 1.4). Subsequent analyses showed that many of the gas-rich meteorites were, physically and chemically, the least altered meteorites, and their gases therefore appeared to be primordial.

1.5.5. Instrumentation and the Discovery of Elements and Minerals

Figure 1.6 summarises the growth of our knowledge of the mineralogy and chemistry of meteorites. The elements seem to have been discovered at a steady, almost exponential, rate. The important factors in this growth were spectroscopic and radioactive methods of analysis, both of which are in routine use today.

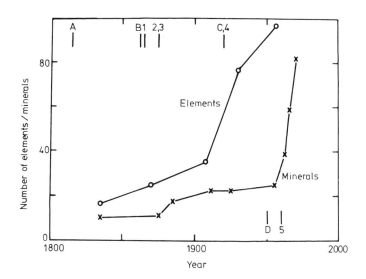

Figure 1.6. Rate of growth in the number of elements and minerals known to occur in meteorites. The letters and numbers indicate the introduction of techniques relating to the identification of elements and minerals, respectively. These are: A, wet chemical analysis perfected; B, qualitative spectroscopy; C, quantitative spectroscopy; D, radioactivity methods: 1, thin sections; 2, polarising microscope; 3, reflected light microscope; 4, x-ray diffraction; 5, electron microprobe.

22

The use of radioactivity in analysis takes two main forms. The most straightforward methods utilise natural radio-activity: the amount reflects the abundance of the nuclide, and its identity can be determined from the rate of decay. This method is obviously limited to the elements which have radioactive isotopes. Instrumental neutron activation over-comes this problem, by making radioactive products by neutron bombardment of the specimen. For increased precision, a chemical separation may be used (radiochemical methods). The main instrument for isotope analysis is the mass spectrometer; this identifies the isotope from its mass-to-charge ratio which is measured by its deflection in magnetic and electric fields. It may be further linked to chromatographic apparatus, which separates out the constituents of a mixture according to their solubility, or absorption properties, on passing through suitable material. This linked equipment provides the standard method of analysis of organic constituents.

The identification of minerals has advanced in an approximately stepwise fashion, the steps being due to (i) the introduction of the microscope, first with thin sections and then with polished sections using reflected light; (ii) the discovery of x-ray diffraction, which enables mineral identification from crystallographic dimensions, and (iii) the electron microprobe. The electron microprobe has changed our knowledge of the mineralogy of meteorites more than any other instrument. An electron beam of about 20 keV energy is focused onto a polished surface of the specimen, causing each element in the specimen to emit x-rays of a characteristic energy, with the intensity directly related to the amount of that element present. Using standards for calibration, this technique enables the chemistry of grains less than 10 μm in diameter to be determined. Its application to meteorites has resulted in the discovery of about 50 minerals. In addition, measurements on the major minerals have exposed inhomogeneities, whose significance are major themes of current meteorite research. The present list of known minerals amounts to somewhere near 100 but, fortunately, most of them are present, for our purposes, in insignificant amounts.

2. Meteorite Falls and Associated Phenomena

2.1. Introduction

The velocity against height curve for a meteorite travelling through the Earth's atmosphere is calculated by numerical solution of the classical aerodynamic drag equation

$$\frac{\mathrm{d}v}{\mathrm{d}t} = -C_{\mathrm{D}}\rho v^2 \frac{A}{m} + g \cos\theta, \qquad (2.1)$$

where v is the velocity, m is the mass and A is the cross sectional area of the meteorite; t is time, C_{D} is the drag coefficient (which depends on shape, but is usually about $2 \cdot 0$), g is the acceleration due to gravity, and θ is the inclination of the path of the meteorite to the vertical. An estimate must also be made of the density of the atmosphere, ρ, at various heights. The resulting curve (figure 2.1) shows initially an exponential decrease in velocity, reflecting the exponential increase in atmospheric density in the first term. At a relatively low height, there is an inflexion in the curve, as the first term becomes less important than the second. This is sometimes termed the 'retardation point'. It is as if the 'cosmic velocity' has been lost and the body falls under gravity, as any body would fall if dropped from the height concerned (Haidinger 1869, Heide 1964, Baldwin and Shaeffer 1971). As a consequence, the impact velocity of most meteorites is very low, about $100-200 \mathrm{~m~s}^{-1}$. This is sufficient to make a hole a few feet deep in most soft soils. The retardation point is suppressed slightly if the meteorite has a lower velocity when it enters the atmosphere, since this lowers the drag.

The change in velocity with height for meteorites of differing mass is also shown in figure 2.1. With larger masses, the retardation point is lowered, until eventually the meteorite

24

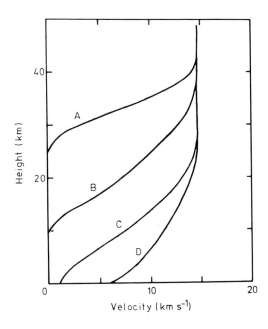

Figure 2.1. Theoretical height against velocity curves for stony meteorites with a 14·2 km s^{-1} entry velocity and inclined 52° to the vertical. A, 500 kg C chondrite (low-density stony meteorite); B, 500 kg H chondrite (high-density stony meteorite); C, 1000 kg H chondrite. In calculating curves A–C, the effect of break-up has been included. D, 1000 kg H chondrite without break-up. This probably approximates most closely to the curve for iron meteorites (from Baldwin and Shaeffer 1971, by permission of NASA Ames Research Center).

impacts the Earth before this point is reached. The meteorite is then required to lose its kinetic energy instantaneously: an explosive impact and a crater result. Krinov (1966) distinguishes between 'explosive craters', 'impact craters', and 'impact holes'. The first correspond to major craters, where essentially complete vaporisation of the meteorite occurs. Where meteorite fragments are occasionally found near such craters, they appear to have become detached from the meteorite before impact. Impact craters and holes are non-explosive but, in the former case, the meteorite is destroyed mechanically, whereas in the latter, large specimens are usually found. For the major Sikhote Alin fall, where more than 100 craters and holes were made, Krinov has been able to

25

show that the divisions may be made at 20 and 9 feet, but it is doubtful that these figures are generally applicable.

Baldwin and Shaeffer (1971) have extended the calculations by allowing for fragmentation and ablation. The body is assumed to break up when the force on it exceeds its compressive, or tensional, strength. This only occurs to any extent with the highly friable C chondrites. In fact, over a certain range of initial velocities the final mass recovered is greater when fragmentation has occurred, than if the meteorite had remained as a single body. This is because the fragments suffer greater drag and ablate less. The rate of ablation (dm/dt) is given by

$$\frac{dm}{dt} = -\tfrac{1}{2}C_H\rho v^3 \frac{A}{\gamma}, \tag{2.2}$$

where C_H is the heat transfer coefficient, and γ is the heat required to ablate 1 kg of material. The results show that, when ablation is allowed for, meteorites of any mass entering the Earth's atmosphere with a velocity in excess of $30\ \mathrm{km\ s^{-1}}$ suffer more than 99% ablation. This seems to be consistent with what is known about the entry velocities of meteorites. The Pribram (Czechoslovakia), Lost City (Oklahoma) and Innisfree (Alberta) meteorites had entry velocities of 20·9, 14·2 and $14\cdot2\ \mathrm{km\ s^{-1}}$, respectively. These three meteorites fell within areas covered by networks of cameras in Eastern Europe and North America set up to photograph the trails of bright meteors and fireballs (McCrosky and Ceplecha 1969, Halliday *et al* 1977). Millman (1969) has compiled data from eye-witness accounts of the fall of 24 meteorites, which give an average velocity of $20\cdot4\ \mathrm{km\ s^{-1}}$ (although errors are frequently about 20%). However, in only one case does the entry velocity exceed $30\ \mathrm{km\ s^{-1}}$.

2.2. Non-Crater-Forming Meteorite Falls

2.2.1. *The Phenomena of Meteorite Falls*

Every meteorite fall has its own character. It will therefore be helpful, before attempting to generalise, to reproduce two

26

eye-witness accounts which emphasise different points. The Limerick meteorite (Higgins 1818) fell at 9.0 am on 10 September 1813:

> ... a cloud appeared in the east, and very soon after I heard eleven distinct reports, appearing to proceed thence, somewhat resembling the continued discharge of heavy artillery. Immediately after this followed a considerable noise not unlike the beating of a large drum, which was succeeded by an uproar resembling the continued discharge of musketry in a line. The sky above the place whence this noise appeared to issue became darkened and very much disturbed, making a hissing noise, and from thence appeared to issue with great violence different masses of matter.... One of these was observed to descend; it fell to the earth, and sank into it a foot and a half.... It was immediately dug up, and I have been informed by those that were present, on whom I could rely, that it was then warm, and had a sulphurous smell.

The most notable feature of this account are the noises. By contrast, when the Barbotan meteorite fell at 10 pm on 24 July 1790 it was the light phenomena which were most impressive.

> ... we found ourselves surrounded, all of a sudden, by a whitish clear light, which obscured that of the [nearly full] moon, ... On looking upwards we observed, almost in our zenith, a fire-ball of larger diameter than the moon. It had behind it a tail, the length of which seemed to be equal to about five or six times the diameter of the body; at the place where it was connected with the body it had about the same breadth, and decreased gradually till it ended in a point. The ball and tail were of a pale white colour; but the point of the latter was almost red as blood....
> Scarcely had we looked at it for two seconds when it divided into several portions of considerable size, which we saw fall in different directions, and almost with the same appearance as the bursting of a bomb. All these different fragments became extinguished in the air, and some of them, in falling, assumed that blood-red colour which I had observed in the point of the tail.
> About three, or perhaps two minutes and a half later ... we heard a dreadful clap of thunder, or rather explosion as if several large pieces of ordnance had been fired off together.... Some time after, when it had ceased, we heard a hollow noise, which seemed to roll along the chain of the Pyrenees, in echoes, for the distance of fifteen miles; ... and at the same time we perceived a strong smell of sulphur.
> ... we observed a small whitish cloud.

The description of the fall of the Barbotan meteorite is not unlike a painting of the Ochansk meteorite fall (figure 2.2). In this account, by Baudin (1798), the author mentions windows being thrown open and kitchen utensils being thrown from shelves, as if during an earthquake.

Figure 2.2. A painting of the fall of the Ochansk meteorite at 12.03 pm on 20 August 1847 (Farrington 1915).

Light and sound phenomena appear to be the most frequently mentioned features of meteorite falls. Table 2.1 lists some of the major features of the falls of 20 meteorites and gives the number of times they are mentioned. There is usually an explosion followed by a series of noises, most frequently described as rumblings, and there is often whistling. The next most frequently mentioned effect is the light produced by the fireball. All meteorites must produce a fireball of some kind: the low figure in table 2.1 indicates that some are not bright enough to attract attention in full daylight. As we shall see, however, some are so bright as to compete with the Sun. The third major feature of the fall of a meteorite is the dust trail left in the sky after the fireball.

Table 2.1. Features mentioned in the fall descriptions of 20 meteorites in the *Meteoritical Bulletin* and their frequency of occurrence.

Explosion	17 (85%)	Light	11 (55%)
Rumbling	7 (35%)	Flares	2 (10%)
Whistling	9 (45%)	Dust trail	6 (30%)
Impact sounds	6 (30%)		

(i) *Sound Phenomena.* The sound phenomena seem to be of three kinds: an initial explosion or set of explosions; a series of smaller explosions, frequently described as rumblings; and whistling or hissing noises. The timing of these events is most important. As we see from the Barbotan account, the explosions and the rumblings which follow occur some time after the fireball is seen, because sound takes much longer than light to reach the observer. These sounds are explained in terms of the mechanics of waves in the atmosphere. Wind-tunnel experiments illustrate how shock waves set up in the atmosphere by the passage of an object will be heard as a sonic boom. The initial explosion heard during a meteorite fall is probably the sonic boom of the leading mass, which has undergone little or no fragmentation. A set of explosions results if there are several major fragments. The experiments also illustrate that several shock waves can be set up when considerable fragmentation occurs. Such a complex series of shock waves would result in the rumbling noise that follows the first major fragmentation (Heide 1964).

The third type of noise—whistling or hissing—is more difficult to understand. The major problem is that the noise is always heard simultaneously with the fireball being seen, as if the sound had found a way of travelling at the speed of light. Some authors believe these sounds are imagined by observers, but Astopovich and, more recently, Romig and Lamar (1963), have compiled catalogues of these 'electrophonic' noises. It may well be that the effect is real. Possible explanations are that the sounds are due to electrostatic discharges near the observer, caused by the creation of static electricity by the fireball, or that they are due to the effect of the fireball's electromagnetic radiation on the human ear. During the fall of the Peace River meteorite, hissing noises were heard at the same time that reception on police radios became difficult (Folinsbee and Bayrock 1964).

In the past few years, it has been possible to detect instrumentally the sound waves produced by meteorites in the atmosphere. If intense enough, the sound waves are communicated to the ground and can be detected by seismometers. Such was the case of the Vilna meteorite which was detected by the University of Alberta seismometer station at Edmonton (Folinsbee *et al* 1969). Although the Tunguska event was detected seismically as long ago as 1908, it is only recently that the equipment has become sensitive enough to be used with normal fireballs. These studies are therefore still at an early stage.

The size of the region over which sounds can be heard is highly variable, depending on weather conditions and on the height at which the sounds are produced. The latter depends on a number of factors; for example, the angle at which the meteorite enters the atmosphere, and the velocity of the meteorite. Vilna, which produced a very small amount of material (totalling 140 mg), but considerable effects in the atmosphere, was heard at a distance of 70 km; Peace River (which produced 49·3 kg) was heard over a 4000 mile square area; Lost City (which produced 17 kg) was heard over 300 km^2, and Barwell (47 kg) was heard over an area 50 km wide and 150 km in length along the track of the fireball.

(ii) *Light Phenomena.* Many accounts, such as that of the Vilna fall, describe the fireball as resembling a comet with a circular head and a long tail. One of the most vivid and detailed accounts is that reproduced for the Barbotan fireball. In this account, a terminal burst is also mentioned, and this is very common. It represents the stage when the meteorite reaches the retardation point, after which it is no longer luminous. In several cases where the head of the fireball has been well examined (for example, Bovedy), multiple centres have been seen, which fall away as separate globules.

Several fireballs have been photographed, so that it is possible to calculate reliable 'light curves' for them. Light curves are plots of the stellar magnitude of the fireball (M) against time, and such curves carry a wealth of information. The width of the trail on a photographic plate is proportional to the intensity (I), so that magnitude can be calculated from $M = -2 \cdot 5 \lg I$. Figure 2.3 shows two light curves for meteorites with very different recovered masses. Vilna (140 mg) produced a fireball which pulsed in intensity, presumably as it rotated (Folinsbee *et al* 1969). It lasted for 7 s, reaching a maximum magnitude of -9 after 4 s, and ended in a terminal burst of -10. A feature of the Vilna fall was a shower of red sparks, produced after the terminal explosion. The recovered mass of the Lost City meteorite (McCrosky *et al* 1971) was 17 kg, and its fireball intensity was much greater than Vilna, reaching a maximum of nearly -12 after 6 s. The intensity of the fireball rarely varies as steadily as in the case of Lost City. Usually there are small flares along the main flight path, and sparks continually fly out.

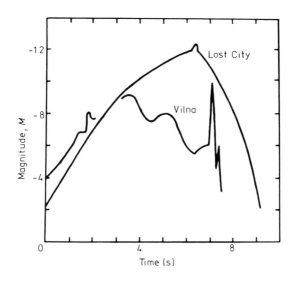

Figure 2.3. Light curves of two fireballs associated with recovered meteorites. (Lost City curve from McCrosky *et al* 1971, Vilna curve from Folinsbee *et al* 1969.)

The intensity of a fireball is related to the rate at which the meteorite loses its kinetic energy ($\frac{1}{2}mv^2$). A certain proportion, τ (termed the 'luminous efficiency'), is converted into light, and the rest to heat and sound, so that we may write

$$I = -\frac{\tau}{2}v^2\frac{dm}{dt},\qquad(2.3)$$

where v is the meteorite's velocity, m is its mass and t is time. τ is a complicated function of velocity and composition. It is frequently assumed that $\tau = \tau'v$, where τ' is determined from artificial meteors launched from sounding rockets. Attempts have been made to combine equation (2.3) with the light curves to calculate the mass consumed during atmospheric flight, but τ is not yet sufficiently well known to yield reliable results (McCrosky *et al* 1971).

The light emitted from a fireball comes from two sources. Material ablated from the meteorite may become heated enough to produce line emission. Two elements are thought to be efficient and abundant enough to be important in this process: iron and sodium. The second source of light is the enormous volume of heated and ionised air, the 'gas cap', which is the object actually observed. Heide (1964) gives a graphic illustration of the relative sizes of the meteorite and gas cap by likening them to a football and a football pitch. In this case, the elements which contribute to the light are, of course, oxygen and nitrogen.

The colour of fireballs varies. Many (such as Barwell) appear to be yellow because of their sodium content. Frequently, the dominant impression is green (such as Vilna). Faint objects often appear to be green because the human eye is most sensitive to it, but fireballs are so bright that it must be a real effect, presumably due to ionised nitrogen. The Peace River fireball was described as being dark orange, or red.

(iii) *Dust Trail.* The mass of the meteorite is reduced considerably throughout its atmospheric passage: liquid drop-lets from the molten surface are blown away and material is evaporated. This gives a continuous stream of fine dust along the path of the fireball, and in some instances a trail is seen. Later this may bend and break up, as motions in the upper atmosphere take effect. According to Krinov (1960), the trail may thicken into a 'cloudlet' at the bottom: this may be the cause of the clouds described in historical accounts of meteorite falls. Attempts have been made to collect this dust from a fireball's trail by flying aircraft down the path of two fireballs (Carr 1970). Revelstoke, which produced very little recoverable material, caused much dust to be present in the atmosphere; while Allende, which is one of the largest meteorite falls known, left very little material in the atmosphere. It is not clear whether such differences are due to the texture of the objects, or their entry velocities.

(iv) *Other Effects.* In addition to the light and sound effects, fireballs may also produce a smell. This is a constant feature of early reports, and has occasionally been mentioned in recent accounts. The smell is usually described as sulphurous: other terms are only occasionaly used (such as onion-like). The same smell lingers with the stone for hours (for example, Crumlin) or perhaps days after the fall, so it is not likely to be imaginary. It is possibly caused by the burning of sulphides in the meteorite, or by the reaction of the ionised air which surrounds the fireball to produce ozone. In either case, it is possible that the gas becomes sealed into the porous meteorite mass when the surface crust solidifies, and then diffuses out subsequently.

2.2.2. The Scatter Ellipse

When a meteorite reaches the surface of the Earth it only rarely remains intact. Normally it has fragmented and, if many pieces have reached the surface, they will be showered over a wide area. Meteorite showers are therefore very common. The largest was probably Pultusk, which was composed of some 100 000 fragments. Holbrook produced 14 000 fragments; L'Aigle about 3000; Tenham and Sikhote Alin more than 300 each.

The fragments do not distribute themselves randomly over the impact site. As might be expected, they are found over an ellipse (the 'dispersion' or 'scatter' ellipse) elongated along the direction of the flight path. The larger fragments are almost always found at that end of the ellipse towards which the meteorite is travelling. This is because the kinetic energy of the fragments is greater for larger masses, since every fragment originally had the same velocity. The heavier fragments require more breaking, and travel further, before arcing down under gravitational pull to Earth.

2.2.3. Morphology and the Fusion Crust

(i) *Morphology*. Most meteorites are simply rock fragments with no particular shape. A significant number, however, are conical (figures 2.4 and 2.5). In these cases, the atmosphere has worn away the corners of the meteorite as it travelled apex first. These meteorites are referred to as 'oriented', because they must have maintained the same orientation throughout a major part of their atmospheric flight. Nininger (1936) has pointed out that orientation is much more common amongst iron meteorites, presumably because of their smaller tendency towards fragmentation. It is also more common for meteorites which fall in the morning to be oriented than those which fall in the afternoon. Conical meteorites come in various sizes. The Willamette iron (figure 2.5)† and the Long Island stone

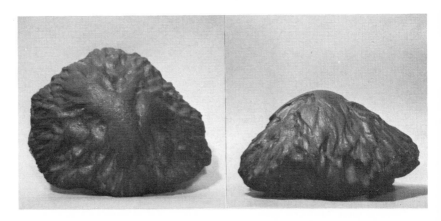

Figure 2.4. The Middlesbrough stony meteorite. Middlesbrough is a very fine example of an oriented, and therefore conical, stony meteorite. It measures about 15 cm across the base and is about 7 cm high. (Yorkshire Museum photographs.)

---

† The Willamette meteorite was the subject of a famous United States court case. The poor farmer who found the iron saw his way to a fortune by selling it to one of the major museums. After considerable effort, he moved it through a forest to his farm. Following the ensuing court case, he had to surrender the meteorite to the owners of the land on which it was found, who then sold it for $26 000. The buyer donated it to the American Museum of Natural History where it now resides.

Figure 2.5. The Willamette meteorite being moved through a clearing on a specially made trolley. The meteorite weighs $13 \cdot 5 \times 10^3$ kg, and measures about 3 m across the base.

are very large, whilst others, such as the Boogaldi iron and the Middlesbrough stone (figure 2.4), can be held in one hand. Some of the oriented Estherville 'pellets' are less than 2 cm in diameter.

There are also a few meteorites with unusual, and highly perplexing shapes. Babb's Mill iron meteorite is barrel-shaped, one metre long and about 30 cm in diameter. One of the Tucson irons has a hole in it up to 67 cm in diameter and is known as the 'Ring Iron'. The Algoma iron is a large flat sheet of metal which measures $25 \times 15 \times 2 \cdot 5$ cm.

On faces which are reasonably large, say more than 10 cm across, it is sometimes possible to see semi-regular depressions resembling thumbprints in soft clay ('regmaglypts'). Their cause is uncertain, but they resemble depressions made in blocks of rock salt after they have been pulled through water behind a motor boat. It is possible, therefore, that they are caused by the motion of the air around the meteorite. There is

35

certainly a wave-like appearance to the regmaglypts on one of the Barwell fragments.

(ii) *The Fusion Crust.* Perhaps the most convincing evidence of the meteorite's passage through the atmosphere is its surface. The heat generated during atmospheric passage melts the surface of the meteorite, and the sudden cooling which takes place when the retardation point is reached produces the fusion crust. It is remarkably thin, usually only 0·3–0·5 mm in thickness, but occasionally, where it has formed a pool in protected depressions, it may be 2–3 mm thick. On most meteorites the fusion crust is black. The texture of the crust is governed by the range of temperatures over which the minerals composing the meteorite melt. Most melt over a range of 1180–1410 °C (Wood 1963, Ivanova *et al* 1968), so that the crust will be sufficiently mobile to flow and ablate while some of its material is still partially solid. The result is a dull, rough crust. In contrast, the minerals in some achondrites appear to melt over a small range of temperatures, so that the material is wholly melted and flows readily to produce a smooth glossy crust. On a few almost iron-free meteorites, this crust may even be transparent.

The molten material on the surface of the meteorite will clearly be violently blown about by the atmospheric motion. The flow patterns formed on the crust are therefore indicative of the orientation of the surfaces on which they are found. The front is smooth and virtually featureless; the sides show long thin streamers of material which has been blown tangentially; and the rear has a cindery, warty appearance (figure 2.6). A classification scheme for fusion crusts has been developed by Krinov (1960): classes I, II, and III, represent front, lateral and rear faces, respectively, and 1, 2 and 3 represent various textures. Crusts of class I1 are thus frontal and smooth, while fusion crusts of class III2 are scoriaceous rear faces. The scheme is not widely used, however, and intermediate textures obviously occur.

Various other types of marking are also seen on the fusion crust; for example, spots resembling oil patches on paper, and large (4–5 cm) patterns resembling comets. The spots are truncated chondrules, and the comet-like marks may be the

(a)

(b)

Figure 2.6. A specimen of the Barwell meteorite showing fusion crust faces with two types of texture. (*a*) Long, thin streamers show that during flight this face was parallel to the direction of travel. (*b*) This face has a cindery, warty texture and was at the rear of the specimen in flight. The other faces were produced by fragmentation after the crust had formed. (British Museum specimen BM 1966 57.)

result of either splashes from an object hitting the molten crust, or flow patterns around a particular protuberance (possibly a large piece of metal).

Under the microscope, the fusion crust is seen to consist of three zones. The material which actually flowed is a black opaque glass, containing sub-micrometre-sized inclusions of magnetite. Beneath this, the silicate grains have melted around the edges, so that they are embedded in the glass. The third zone begins abruptly where the temperature reached 1100 °C. Here sulphide–metal eutectics formed, and these have readily flowed further into the meteorite. The distance, y, of the eutectics from the surface of the meteorite is governed by the rate of ablation, v_w, and is greater for small ablation rates. The equation

$$v_w = \frac{K}{-\rho C_p} \frac{1}{y} \ln\left(\frac{T}{T_i}\right) \tag{2.4}$$

where K is thermal conductivity, ρ is density, T_i is the surface temperature and T is 1100 °C, can be used to calculate ablation rates. It gives rates of 0·35, 0·22 and 0·18 cm s^{-1} for the front, side and rear faces, respectively, of a typical meteorite (Sears and Mills 1973).

The fusion crust of iron meteorites consists of several coatings of magnetite and metal with no silicate glass. The layers are thought to be the result of several coatings of melted material flowing over one another. The structure of iron meteorites is very temperature-sensitive, and a 'heat alteration zone' of about 1 cm in depth exists around the edges. Inside this zone only a granular texture exists, the hardness of which is related to the temperature experienced. Again, it is possible to determine the temperature at a given depth and to apply equation (2.4) to find the ablation rate. Values of about 0·2 cm s^{-1} are obtained (Lovering *et al* 1960). The higher thermal conductivity of the irons, compared with stony meteorites, appears to be balanced by their higher density.

2.2.4. Meteorite Recovery

There appear to be variations in both the seasonal and diurnal fall rate of meteorites. More falls occur in the period April to July than during November to March, but this may simply reflect the fact that most observations are made in the northern hemisphere and that more people are out of doors during the summer months to witness the falls. Seasonal variations in the fall rate have many times been used to seek relationships between metorites and meteors, usually with negative results.

The diurnal variations are illustrated in figure 6.3. There is a marked increase in the number of falls around late afternoon, whilst a minimum is found 12 hours earlier. This suggests that meteorites may tend to reach the Earth on a particular kind of orbit: this possibility will be discussed in greater detail below (§ 6.3.4).

The actual fall rate for meteorites over the entire Earth is difficult to determine with any certainty. Only a small fraction of falls are observed and reported, as most of the Earth's surface is covered with ocean, or is sparsely populated. The general level of education of the population in an area also seems to affect the number of meteorite falls reported. Using figures from the densely populated regions of India, Brown (1960a,b) estimated that one fall per year occurred over $10^6 \, km^2$. Over the Earth as a whole, this corresponds to 560 meteorites per year. Hawkins (1960) carried out similar calculations using figures for western Europe and eastern USA, and found that 1200 meteorites of greater than 10 kg fell per year. In contrast, the museums of the world probably receive about six falls per year!

About half of the meteorites in our collections were actually seen to fall, and are referred to as 'falls'. The remainder were found and recognised as meteorites and are therefore called 'finds'. The distinction is not as trivial as it may seem, because many properties are affected if a meteorite has lain on the surface of the Earth for hundreds, or thousands, of years. For example, there would be little point in looking for water-soluble minerals in finds. The exact numbers of falls and finds for irons, stony-irons and stones, according to Hutchison et al (1977), are given in table 2.2. These figures show that stony meteorites fall much more frequently than irons, but irons still constitute a large proportion of the collections because they are found more easily. This is presumably because irons resist weathering better than stones, and so stand a better chance of being found before they are destroyed. In addition, irons are

Table 2.2. Numbers of meteorite falls and finds of stones, stony-irons and irons (Hutchison *et al* 1977).

	Falls	Finds	Total
Stones	791	587	1378
Stony-irons	11	66	77
Irons	46	595	641
Total	848	1248	2096

comparatively easy to recognise. Their fusion crust, high density (about $8\,\mathrm{g\,cm^{-3}}$, compared with about $3\,\mathrm{g\,cm^{-3}}$ for most terrestrial rocks), magnetism, and ringing tone when struck with a piece of metal, are all characteristic. Stones are less readily recognised. When comparatively fresh, a fusion crust with flow marks is probably their best recognition feature. The majority of stony meteorites contain small irregular grains of metal, so that most meteorites are magnetic to some extent. Under a magnifying glass, stony meteorites often contain rounded blebs of silicate called chondrules. These are not always readily discernible, however, and frequently only specialists can make a definite identification.

Figure 2.7 shows the location of all the finds of meteorites listed in the *Catalogue of Meteorites* (Hey 1966). A number of features illustrate some of the factors involved in recovering finds. Most conspicuous is the large number of finds in the American Prairie states, where most of the stony meteorite finds have been made. This is probably a consequence of the history of the region. Its climate has favoured preservation, and the soil colour and land use have aided recovery of meteorites; but the major factor has been the sudden intensive inhabitation and ploughing of areas which had existed for thousands of years as grassland. The large number of finds in Australia is probably the result of similar factors. The Chilean finds present a special problem, since they are predominantly iron meteorites of one structural class (hexahedrites): 16 have been found in this region, mainly in Chilean saltpetre quarries, whereas over the rest of the world only 38 have been found and six seen to fall. At first sight, it seems most probable that they are a single major fall and, indeed, all but one

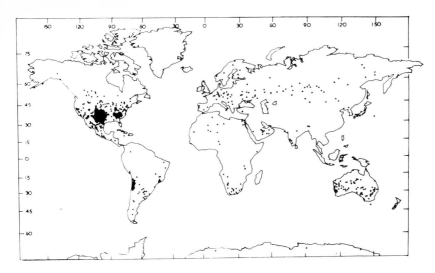

Figure 2.7. Map showing the location of all sites where meteorites have been found which were not observed to fall (data from Hey 1966).

(Tamarugal) have similar terrestrial ages (although the error limits are very large). Trace element chemistry suggests, however, that at least four falls are present (Wasson and Goldstein 1968).

The line of finds across Asia marks rather well the route of the trans-Siberian railway. Since townships have been established along its length, this also reflects the population distribution after the construction of the railway. Regions such as western Europe, which have always been reasonably well populated, are deficient in finds, since most meteorite falls in this region were witnessed. In recent years, the Antarctic has proved to be a remarkably efficient collector of meteorites. This is because they tend to concentrate in regions where evaporation of the ice occurs. Since 1969 the Japanese have found several hundred meteorites around the Yamato (Queen Fabiola) mountains, and in less than two months a joint US–Japanese team found eleven meteorites near McMurdo Base (Cassidy *et al* 1977).

2.2.5. Terrestrial Ages of Meteorite Finds

The terrestrial ages of meteorites may be determined by radioactivity methods. Exposure to cosmic radiation induces

41

nuclear reactions, some products of which are radioactive
(§ 5.3). When the meteorites reach the Earth, the production
of these radionuclides ceases, and their activity begins to
decay. If the decay rate and the original activity are known,
then the time that has elapsed since they fell—that is, the
terrestrial age—can be calculated. For iron meteorites, the
nuclides which have been used, and their half-lives, are ^{36}Cl
($3 \cdot 08 \times 10^5$ years) ^{39}Ar (325 years) and ^{10}Be ($2 \cdot 7 \times 10^6$ years),
while for stony meteorites, ^{14}C (5760 years) is the most suit-
able. The choice of nuclide depends on its abundance in the
meteorite and the age expected. Stony meteorite terrestrial
ages are typically a few thousand years: for iron meteorites,
this figure is frequently of the order of 10^5 years (Anders
1963, Begemann and Vilcsek 1969, Begemann and Wänke
1969, Chang and Wänke 1969, Boeckl 1972). The major
cause of error in the calculation of age is the original activity,
which has to be measured in several recent falls and then
assumed constant for all meteorites. This is not a very good
assumption, and errors in terrestrial age measurements are
usually very high. The estimation of production rates is discussed in § 5.3. A possible additional technique for some
stony meteorites, which has not yet been fully exploited, is
thermoluminescence.

2.3. Explosive Meteorite Craters

2.3.1. Impact Mechanism of Giant Meteorites

The mechanism of meteorite impact can be studied by simulation experiments and by theoretical methods. Öpik pointed
out in 1936 that, when a meteorite strikes the Earth with
cosmic velocities, the elastic limits of both meteorite and
target are exceeded, and both act as fluids. This enables
hydrodynamic equations to be applied in determining the
behaviour of projectile and target. There are four equations
which specify the problem and require solution. The first is
the equation of motion:

$$\rho \frac{\mathrm{d}v}{\mathrm{d}t} + \nabla p + \rho \nabla V = 0, \qquad (2.5)$$

where ρ is the density, v is the projectile velocity, and V is the potential function. The second equation expresses conservation of mass, and may be written as:

$$\rho V \cdot \mathbf{v} + \frac{\mathrm{d}p}{\mathrm{d}t} = 0. \tag{2.6}$$

The third is the energy conservation equation (assuming adiabatic conditions):

$$\frac{\mathrm{d}E}{\mathrm{d}t} = \frac{p}{\rho^2} \frac{\mathrm{d}p}{\mathrm{d}t}, \tag{2.7}$$

where E is the energy of the projectile, and p is the pressure set up in the target material. A final equation of state is also required, relating the density of the target projectile and the energy of impact in a form analogous to that of the ideal gas equation. This last is a difficult function to establish, and elaborate semi-empirical relations have been developed. A discussion of these four equations may be found in Kaula (1968), Beals *et al* (1963) and Shoemaker (1963). The results of a numerical solution of the equations are shown in figure 2.8.

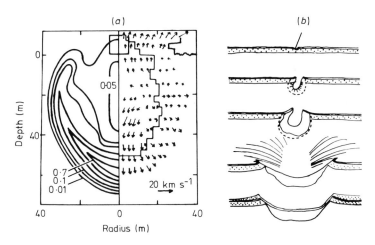

Figure 2.8. Impact mechanism of giant (crater-forming) meteorites: (*a*) pressure–velocity field 3·44 ms after impact of a 12 000 ton iron meteorite with an initial velocity of 30 km s^{-1} (Bjork 1961): the contours refer to pressure in megabars. (*b*) Formation of the Canyon Diablo meteorite crater (Shoemaker 1963).

Taken together, the simulation experiments (Gault *et al* 1968) and the calculations give the following impression of a hypervelocity meteorite impact. Upon impact, a fine spray of material shoots out sideways (the 'base surge') and two shock waves radiate from the point of impact, one into the ground (compressing it), and one into the projectile (decelerating it). For a 12 000 ton iron projectile, moving at 30 km s^{-1}, pressures of 10 Mbar may be experienced at its forward edge, 0·17 ms after impact. A crater opens up in response to the compression, and some projectile material may jet out. Between the opening crater and the shock wave, the ground becomes very hot and may even fuse. After 24·8 ms, the maximum pressure will have fallen to 40 kbar, and a penetration of 90 m achieved. The shock wave in the projectile results in a 'stagnation plane', where it has completely stopped. Below this it is still pressing into the ground, but above it is shooting upwards. The upward motion can be used to calculate the crater diameter and the dimensions of the 'ejecta blanket', the region over which material is thrown out from the crater. In this example, the crater is 1 km in diameter and 150 m deep. The explosion will overturn, and perhaps break up, the strata, so they appear in reverse order in the crater rim.

Nuclear explosion craters bear a striking resemblance to meteorite craters, since in both cases the force producing them is explosive: this is in contrast with the comparatively passive mechanisms which produce volcanic craters. Studies of craters formed by nuclear explosions have led to the definition of a 'radius of crushing' (R) around and under the crater, within which mechanical mixtures of rocks (breccias) are found. Vaporisation seems to occur up to $0·05\,R$, melting to $0·07\,R$, and fracturing up to $2\,R$ from the centre of the explosion. R can be related to the energy of the explosion by the expression:

$$R = 5·7\ W^{1/3}, \tag{2.8}$$

where W is the energy in equivalent tons of TNT (1 ton = $4·16 \times 10^6$ erg). Similarly, the diameter of the crater itself is proportional to $W^{1/3}$.

The most recent estimates for the mass and size of the object responsible for the Canyon Diablo crater (figure 1.4) have been made using expression (2.8). The ratio of the diameter of the Canyon Diablo crater to the Teapot Ess nuclear explosion crater is between 10·5 and 11·4 (depending on how the diameters are measured). The energy of the Teapot Ess explosion is obviously very accurately known, so we can estimate that the energy of the Canyon Diablo impact was 1·4–1·8 equivalent megatons of TNT. From the considerations of the previous section, the velocity of the meteorite must have been in the range 11·2–30 km s^{-1}. Since $E = \frac{1}{2}mv^2$, this indicates a mass of 63 000–166 000 metric tons; or, for a spherical mass of density 7·85 g cm^{-3} (the density of the recovered fragments), a radius of 24·8 m. Similarly, $R = 5·7\,W^{1/3} = 130$ m, which compares reasonably well with 175 m, the best (though still rather poor) estimate for the observed extent of brecciation.

2.3.2. Criteria for Crater Recognition

The number of craters currently recognised as meteoritic is somewhere near 50, depending on how rigorously the criteria are applied (French 1968). The major characteristic of a meteorite impact crater is probably its shape. It is usually circular with walls which incline steeply on the inside and gently on the outside. The rim stands higher, and the crater bottom lower, than the surrounding ground level. The depth–diameter ratio varies, being slightly greater for smaller craters. Some of the larger craters are identical to lunar craters in their dimensions, whereas craters produced by bombs have larger depth–diameter ratios. Another characteristic of both lunar and terrestrial craters is the uneven ejecta blanket (for example, Canyon Diablo; see figure 1.5).

Geophysical measurements have permitted the identification of many impact craters, notably on the Canadian shield, where several ice ages have destroyed the crater rims (Beals et al 1963). They can also be used to distinguish between meteoritic and volcanic craters. Gravity anomalies describing contours may indicate a crater filled with low-density sediment, whilst seismic tests can detect the breccia surrounding the crater. Magnetic measurements are also sometimes useful in magnetite-rich rocks, since any magnetic alignment may be destroyed by the random nature of the breccia. However, confirmation of a structure may require drilling.

Debris which is recognisable as being due to meteoritic impact, can take a wide variety of forms. We will discuss actual fragments of the meteorite itself in the next section. Breccias and impactite glass are good signs of meteorite impact. Impactite is a wholly or partially fused mixture of country rock, sometimes with some finely dispersed meteoritic constituent. In recent craters, rusted mixtures of meteorite material and rock can be found, and are referred to as 'shale balls'. Perhaps the most unexpected signs of meteorite impact were discovered by Dietz in 1943, and are called 'shatter cones'. These are conical configurations adopted by certain kinds of rock in response to the tremendous pressures they have received. They are indicative of violent intense shock pressures of 20–100 kbar, which can only readily be produced by a meteorite impact. They line up with the apex of the cone pointing towards the centre of impact, and thus provide a means of locating the crater centre when crustal movement has obscured it. Small, but perfect, shatter cones have been produced by simulation experiments.

As may be imagined, meteorite impact has a considerable effect on the microscopic nature of the rocks around the crater. Pressures of 100–300 kbar cause planar features, slip bands and deformation lamellae to appear in minerals such as quartz and feldspar. Pressures of 300–500 kbar cause quartz to assume its high-pressure crystalline forms, coesite and stishovite. These are probably the two most important diagnostic features known (French 1968).

2.3.3. Craters With Meteorite Fragments

A few craters, inevitably the first discovered and the best known, have meteorite fragments associated with them (Krinov 1966). Some of these craters are listed in table 2.3. Most are in dry, desert-like environments such as those in Australia. With the smallest craters, there is usually some doubt as to whether they are the result of hypervelocity impact—and therefore explosive—or are simply impact holes.

Table 2.3. Craters with meteorite fragments (Krinov 1966, Hey 1966, Hutchison *et al* 1977).

Crater name and location	Number	Year discovered	Diameter ×depth (m)[1]	Material recovered
Meteor Crater (Canyon Diablo), Arizona 35° 0'N : 111° 0'W	1	1891	1207 × 174	more than 3 × 10^4 kg
Odessa, Texas 31° 43'N : 102° 25'W	4	1921	168 × 5	more than 22 kg
Henbury, Northern Territory, Australia 24° 34'S : 133° 10'E	13	1931	220–110 × 12–15	more than 1.5×10^3 kg plus impact-ites and stone flour
Wabar, Rub'al Khali, Saudi Arabia 21° 29.5'S : 50° 40'E	2	1932	100 × 12	11·5 kg plus impactites
Haviland (Brenham), Kansas 37° 37'N : 99° 5'W	1	1933	17–11	more than 850 kg plus meteorodes
Campo del Cielo, Gran Chaco, Gualamba, Argentina 27° 28'S : 63° 42'W	2	1933	75	more than 100 kg plus impactites
Kaalijarv, Oesel, Estonia 58° 24'N : 22° 40'E	7	1937	110 × 16	0·6 kg plus stone flour
Boxhole, Central Australia 22° 54'S : 135° 0'E	1	1937	175 × 16	about 90 kg
Dalgaranga, Western Australia 27° 45'S : 117° 5'E	1	1938	21 × 3·2	about 2 kg
Wolf Creek, Western Australia 19° 10'S : 127° 46'E	1	1947	853 × 46	about 64 kg
Sikhote Alin, Eastern Siberia 46° 9·6'N : 134° 39·2'E	27[2]	1947	26·5 × 6·0	more than 2.3×10^4 kg
Monturaqui, Chile 23° 57'S : 68° 17'W	1	1975	370 × ?	iron (IAB) shale

47

Table 2.3.—*cont.*

Crater name and location	Number	Year discovered	Diameter ×depth (m)[1]	Material recovered
Um-Hadid, Saudi Arabia 21° 30′N : 50° 40′E	1	1971	10×?	more than 1 kg and impactite

[1] Where several craters are present the dimensions correspond to the largest.

[2] This is the number of craters greater than 7 m in diameter and which may be considered as explosive craters (Krinov 1966). An additional 131 impact pits and holes were present at the site.

Doubts have been expressed about the explosive origin of the Haviland, Campo del Cielo and Dalgaranga craters, and they will not be discussed at length here. Curious eroded masses were found in the Haviland crater which were said to resemble potatoes, and were termed 'meteorodes' by Nininger. An additional, recently discovered crater exists on Mount Darwin, Tasmania, where a considerable quantity of impactite glass ('Darwin glass') has been found.

The pieces of meteorite associated with impact craters are fragments which either broke away from the meteorite as it travelled through the atmosphere, or were spalled off the back of the meteorite during the first milliseconds of impact. These two types of fragment have been identified around the Canyon Diablo meteorite crater, the discovery of which was described in Chapter 1. The atmospheric passage of Canyon Diablo has been particularly well studied; partly because of its association with the crater, and partly because it is the only iron meteorite found to contain diamonds. The diamond-bearing fragments were all found on the crater rim. They were all badly shocked and, hence, when polished and etched have a characteristic appearance. By contrast, the specimens found on the sur-rounding plain contain no diamonds and are not shocked. Presumably, therefore, the plains specimens broke off from the meteorite in the atmosphere, whilst the rim specimens spalled off the back on impact: the shock which produced the diamonds. Radiochemical analysis has confirmed that the shocked fragments came from the rear of the meteorite (Heymann *et al* 1966). A considerable number of millimetre-sized spherules have been found in the soil: these were presumably formed from the evaporated material immediately after the impact.

Estimates of the age of the crater vary from 900 years—this being the age of a tree in the crater—to 200 000 years. The most generally accepted figure is 50 000 years, as calculated from the extent of erosion of the limestone in which the crater is situated (Krinov 1966). Various legends are said to exist that the North American Indians knew of the impact, but these were not verified by a rigorous study in 1943, and it is doubtful that Arizona was populated by the Indians at the time of the impact.

In 1927, the second meteorite crater to be discovered after Canyon Diablo was found near Odessa in Texas. It is very eroded and probably older than Canyon Diablo; at present it is only 168 m in diameter and 18 m deep. A fossilised horse of an extinct type was found in the crater and confirms its great age. In 1939, a number of impact holes were discovered near the crater in which three or four large iron meteorite fragments were found.

The next crater to be identified was the first of the many Australian craters containing meteorite fragments. The Henbury craters of Central Australia were discovered in 1930 in a way similar to Canyon Diablo. Pieces of iron meteorite were brought in by Aborigines and, when the source of them was tracked down, an impact site consisting of 13 craters was found. The distribution of the craters is similar to the distribution of the fragments of a shower: larger at one end and smaller at the other. The main crater is elliptical, probably two merged into one, measuring 220×110 m, and the others are 80–9 m in size. The Henbury meteorite fragments are noted for their angularity, which may be associated with the violent atmospheric passage suggested by the large number of craters. The similar angularity of the fragments of the Chinga fall has led Krinov (1966) to suggest that they, too, are associated with a crater, but that the climate in the region has not been favourable to its preservation. As with Canyon Diablo, magnetic nickel-rich spherules have been found in the soil around the Henbury craters.

The Boxhole and Wolf Creek craters (found in 1937 and

1947, respectively) are both isolated individuals from which iron meteorites have been recovered. No other debris have been found near Boxhole, but large shale balls have been found near Wolf Creek.

The Waber crater field in Arabia contains four craters, two of which are filled with sand. They were found in 1932 by an archaeologist, H Philby, who was at the time searching for a lost city. Much impactite glass has been found, in addition to the meteorite specimens, and it contains microscopic spherules of two kinds. One type is comparatively large (100 μm), and is analogous to the more common globules found in the soil around the Canyon Diablo and Henbury craters. Presumably, these are actual droplets of melted meteorite. The second type of spherule also found occasionally in the soil from the Canyon Diablo crater, is much smaller and is concentrated near the edges of the impactite. This is assumed to be a condensate from the vapour cloud.

The Kaalijarv field in Estonia (table 2.3) consists of a field of several craters. The largest, 110 m in diameter, was recognised as being meteoritic in 1927. The site is surrounded by cultivated fields, and trees have grown readily on the crater wall and rim, making it conspicuous. In several smaller craters, a curious dolomite powder has been found, which is evidently the result of an explosion, and is termed 'stone flour'.

2.3.4. Craters Without Meteorite Fragments

Table 2.4 lists several of the better known meteorite craters in which no meteoritic material has been found. Their identification depends wholly, therefore, on the criteria described in § 2.3.2. Most of the craters listed are in Canada, where the large Canadian shield, which has suffered very little tectonic movement, provides an ideal region for looking for old craters. More recently, a search of the Siberian platform is proving fruitful. After the discovery of the New Quebec crater (also known as the 'Chubb crater') on aerial photographs in 1950, systematic searches were made which proved very fruitful. It is the age of the craters that ensures there are no meteorite fragments, and also creates the bias towards larger structures (since several glaciations will have destroyed

Table 2.4. Possible hypervelocity craters without associated meteorite fragments (Hey 1966, Hutchison *et al* 1977)†.

Crater name and location	Year discovered	Diameter × depth (m)
Aouellel, Mauritania 20° 15′N : 12° 41′E	1937	250 × 6·5
Illumetsa, Estonia 58° 0′N : 27° 14′E	1960	80 × 9
Brent, Ontario 46° 4′N : 78° 29′W	1951	3700 × 60
Holleford, Ontario 44° 27′N : 76° 38′W	1956	2400 × 30
Deep Bay, Saskatchewan 56° 42′N : 88° 45′W	1957	13 700 × 340
Chubb (New Quebec), Quebec 61° 47′N : 73° 40′W	1950	3200 × 396
Nordlinger Ries, Swabia, Germany 48° 53′N : 10° 37′E	1965	21 × 24
Steineim, Swabia, Germany 48° 36′N : 10° 33′E	1933	2500 × 80
Mount Darwin 42° 15′S : 145° 36′E	1933	1000 × ?
Lake Bosumtwi, Ashanti, Ghana 6° 32′N : 1° 24′W	1931	10 500 × ?
Charlevoix Structure, Quebec 47° 32′N : 70° 18′W	1968	35 000 × ?
Gosses Bluff, Northern Territory 23° 50′S : 132° 18′E	1968	25 000 × ?
Chassenon Structure, Haut-Vienne, France 45° 50′N : 0° 56′E	1969	10 000 × ?
Sudbury, Ontario 46° 20′N : 81° 10′W	1963	100 000 × ?
Vredefort, Orange Free State, South Africa 27° 0′S : 27° 22′E	1936	100 000

† Over 250 structures have, at some time, been claimed to be meteorite craters. This is an abitrarily chosen list of a few of the better known ones.

smaller craters). Even so, the erosion has affected their shape to such an extent that they are sometimes referred to as 'fossil craters'. The Holleford crater in Ontario contains Cambrian sediments, and is therefore greater than 500 million years old.

It has also suffered water erosion from seas thought to have existed 500–1000 million years ago. The meteoritic nature of the Aouellel and Illumetsa craters is well established since, as well as the geophysical and other evidence, impactite glass has been found near both of them

In addition to craters of the kind described above, which can be recognised from aerial photographs, debris and geophysical tests, there are numerous instances of major impact damage although the crater no longer exists. These 'astroblemes' have been identified as being meteoritic, because they contain shatter cones. A possible example is the Vredefort Ring in the South African Transvaal. Most of the structure is covered with sediments up to 260 million years old, and if the feature were meteoritic this is probably the age of the impact. The 'reassembled' crater is 100 km in diameter, which would require an energy equivalent to 1·5 million megatons of TNT, compared with about 2 megatons required for the Canyon Diablo crater. A similar event occurred in the Sudbury Basin of Ontario, where the impact was fierce enough to trigger the flow of magma from great depths; this being the source of the major ore deposits there.

2.4. Tektites

2.4.1. Appearance and Distribution

Tektites are pieces of glass found scattered over certain regions of the Earth. The history of their study has been reviewed by O'Keefe (1976). A wide range of suggestions for their origin have been made (for example, volcanic glass, prehistoric man-made glass, lightning-fused sand, etc), but two main theories appear to be the most feasible (i) that they are splash material ejected when a meteorite hit the Earth, or (ii) that they are similar material thrown out by an impact on the Moon. An invaluable collection of papers on tektites has been compiled by Barnes and Barnes (1973), whose editorial comments make the case for a terrestrial origin. Against this, O'Keefe (1976) presents the arguments for a lunar origin.

Tektites range in size from microscopic to about 20 cm in length, but are generally a few centimetres or so across. Most are black and translucent when held up to the light, and transparent in thin section (when flow patterns can usually be seen). The majority of tektites appear to be the result of liquid globules solidifying in flight, and are generally called 'splash form'. They are shaped like teardrops, dumb-bells, rods, spheres, discs or flanged buttons. The flanged-button type appears to result from oriented flight through the atmosphere (Baker 1958). By contrast, Muong Nong type tektites are blocky and layered, and seem to be parts of solidified silicate puddles (Barnes 1971).

Tektites do not occur uniformly over the Earth, but are concentrated in a few 'strewn fields'. The largest area is that of the Australian tektites (the 'australites') which covers the whole of southern Australia. The philipinites, javanites and indochinites are other important fields in south-east Asia. Major fields are also found in Czechoslovakia (the 'moldavites', after the Moldau River), Africa (the Libyan Desert glass and the Ivory Coast tektites) and the USA (the Georgia tektites and the bediasites, named after the Bedias Indians). A newly discovered field is that of the 'irizites' in central Russia. A number of microscopic tektites have been found in ocean sediment just off the Ivory Coast and in the Indian Ocean around southern Australia.

2.4.2. Physical Properties and Ages

The most readily determined physical properties of tektites are their specific gravities and refractive indices. These properties correlate positively with each other, and negatively with the silica content of the tektites (table 2.5). The magnetic susceptibility of tektites is higher than obsidian (volcanic glass), but their intensity of magnetisation is lower. This seems to indicate that tektites have undergone greater heating than obsidian. The fusion temperatures of tektites range from 1055 to 1450 °C, and their rate of cooling, determined from strain characteristics, is $50 \,°C \, min^{-1}$. All these physical properties testify to the natural origin of tektites. Artificial glasses with melting points as high as this were only available well after the discovery of the first tektites. Enclosed in the tektites are bubbles (which contain atmospheric gases), grains of quartz

Table 2.5. The specific gravity, refractive index, SiO_2 content and ^{40}K–^{40}Ar ages of tektites (Barnes 1967, Florensky *et al* 1977).

	Specific gravity	Refractive index	SiO_2 content	Age (10^6 years)
Australasian tektites[1]	2·41–2·44	1·51–1·59	70·8–73·7	0·68–0·76
North American tektites[2]	2·38	1·495	76·6	33·7–35·0
Moldavites	2·35	1·490	79·0	14·4–14·9
Ivory Coast tektites	—	—	—	1·3
Libyan Desert glass	2·21	1·462	97·6	—
Darwin glass	2·29	1·481	88·0	—
Irghizites (SiO_2 rich)			74	
(SiO_2 poor)			54	< 8

[1] Australites, indochinites, javanites, philliponites, billitonites.
[2] Bediasites, Georgia and Matha's Vineyard tektites.

and coesite, and nickel–iron spherules. These last two are particularly significant as they are also found in association with meteorite impact craters.

The ages of tektites, as determined by the K–Ar method (§ 5.2.3), are listed in table 2.5. Some additional age data is provided by Rb–Sr and charged-particle track methods. The age of the moldavite tektites (15 million years) corresponds rather well with the age of the Ries Basin in Germany, which was probably formed by meteorite impact. Similarly, the Lake Botsomtwi crater in Ghana was formed at the same time as the Ivory Coast tektites, 1·3 million years ago (Gentner *et al* 1963, Schnetzler *et al* 1966). The recently discovered Mount Darwin crater—which was the source of the Mount Darwin glass—has the same age as the australites (Gentner *et al* 1973), but seems too small to be their source. Dietz (1977) has therefore proposed that their source was the Elgytgyn crater in Siberia. Florensky (1976) has suggested that the irizites and the Zhanabshin impact structure were formed simultaneously. The identification of craters of the correct age constitutes important evidence that the tektites originated as splash material from meteorite impacts on the Earth (Barnes 1967).

2.4.3. Composition of Tektites

Tektites are 70–80% SiO_2; the remainder being primarily Al_2O_3, with FeO, MgO, and CaO sometimes being important constituents (table 2.6). Compositionally, they resemble neither the main types of lunar rocks—Maria basalts or upland

Table 2.6. The composition of the australites (%) and the range shown by all tektites (Barnes 1967).

	Australites (32)	All tektites (168)
SiO_2	73·06	68·00–97·58
Al_2O_3	12·23	1·54–17·56
Fe_2O_3	0·60	0–2·25
FeO	4·14	0·23–6·81
MgO	2·04	trace–4·96
CaO	3·38	0–5·10
Na_2O	1·27	0·01–2·46
K_2O	2·20	0–3·76
TiO_2	0·68	trace–1·40
MnO	0·12	trace–0·42

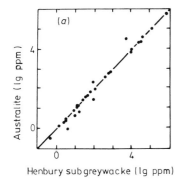 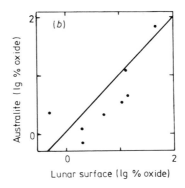

Figure 2.9. Plots comparing the composition of the australite tektites with (a) a terrestrial sedimentary rock (Henbury subgreywacke) and (b) the lunar surface (Taylor and Kaye 1969).

55

gabbros—nor the major oceanic or continental terrestrial rocks. However, it has often been noted that tektite composition does match certain sedimentary rocks (greywacke), especially that from around the Henbury craters (figure 2.9). Unfortunately, none of the meteorite craters thought to be responsible for tektites is situated in this kind of rock.

The absence from tektites of nuclides produced by cosmic ray bombardment rules out any extraterrestrial source but the Moon. There is probably a consensus view that tektites are a form of terrestrial impactite glass, but the details, especially concerning their composition, are far from being understood.

3. Classification, Mineralogy and Petrology

3.1. Introduction

The usefulness of any classification scheme is twofold. Firstly, it provides a descriptive label so that, once a classification scheme has been drawn up, much of what is known about the class may be assumed to apply to any individual in it. Secondly, the establishment of classes may ultimately prove to have genetic relevance since, if a number of meteorites can be put together in one class, it seems very probable that they will have shared a similar origin or history. On the whole, the meteorite classes which have evolved over the last 150 years can be recognised using chemical criteria supported by data which reflect their levels of oxidation. Certain achondrite classes are exceptions, and their original mineralogical and textural definitions are still preferred. Ideally, one would prefer the classes to have distinctly different compositions, although the members of a class should be reasonably similar In many instances, better analyses in recent years have frequently shown that this actually applies. The abundance of volatile elements may be readily affected by subsequent reheating; so for classification purposes, we will concentrate on the less mobile, non-volatile elements. The meteorite classes are listed in table 3.1. In this chapter, we will discuss those classes that are reasonably well populated or of particular interest. Descriptions of the others may be found in the references listed in table 3.1.

3.2. The Chondritic Meteorites

3.2.1. The Chondrite Classes

Some 86% of all meteorite falls are chondrites. The basis for

Table 3.1. The meteorite classes and their populations.

Chondrites	Ca-rich achondrites	Ca-poor achondrites	Stony-iron meteorites	Iron meteorites
E (17)	Eucrites (24)	Ureilites (8)	Mesosiderites (20)	IAB (90)
H (160)	Howardites (19)	Diogenites (8)	Pallasites (41)	IIAB (52)
L (222)		Aubrites (9)		IIIAB (156)
LL (67)				IVA (40)
CB (24)				IVB (11)
CV (11)				
Six anomalous chondrites	Nakhlites (3) Shergottites (2) Angra dos Reis	Chassigny Cumberland Falls	Lodran Steinbach	IC (10), IIC (7), IID (13), IIE (12), IIICD (12), IIIE (8), IIIF (5), 82 anomalous irons

References to descriptions of minor groups and anomalous meteorites not described in detail in the text: nakhlites, Bunch and Reid (1975), Boctor *et al* (1976). Chassigny, Prinz *et al* (1974), Mason *et al* (1976). Angra dos Reis, Angra (1977). Cumberland Falls, Lodran and Steinbach, Mason (1962). Shergottites, Binns (1967). Four anomalous chondrites: Kakangari, Mount Morris (Wisconsin), Pontlyfni and Winona (these may represent a new class which the author terms 'forsterite chondrites'), Graham *et al* (1977). Minor iron meteorite groups, Scott and Wasson (1975).

their classification is shown in figure 3.1, which is a plot of reduced iron (Fe as metal and sulphide) against oxidised iron (Fe in silicates)[†]. The meteorites cluster into several groups[‡]. The H (high iron), L (low iron) and LL (low iron–low metal) chondrites are very similar and are collectively known as 'ordinary chondrites'. On or near to the vertical axis are the E chondrites, named after their dominant mineral, enstatite

[†] Element ratios are used because, unlike percentages, they are not affected by variations in major volatile elements. They are also easier to measure accurately.
[‡] The terms 'groups' and 'classes' are generally considered to be synonymous. The only exception, recently proposed by Wasson (1974), is that while one can put anomalous meteorites together as the 'anomalous class', they cannot be referred to as a group.

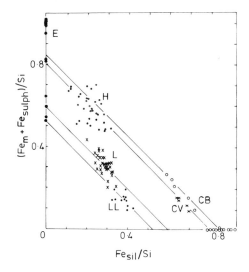

Figure 3.1. Plot of reduced iron (iron in metal and sulphide) against oxidised iron (iron in silicates) for chondritic meteorites. The iron is expressed as an atomic ratio to silicon, and the diagonals indicate constant Fe/Si ratios: from left to right the values are 0·54, 0·59, 0·80 and 0·84. CV and CB chondrites are distinguished primarily by their Ca/Si and Al/Si ratios (figure 3.3). Some further division of the E and CB chondrites is sometimes made. The less oxidised members of CB are termed CO, and the high Fe/Si members of E and the remaining CB chondrites are termed EI, EII and C1, C2, respectively. These are not considered to be different classes as there are no gaps in the Fe/Si distribution, nor in that of any other major non-volatile element. (Data from Müller *et al* 1971, Van Schmus and Hayes 1974, Larimer 1968.)

(MgSiO$_3$). The remaining chondrites are collectively termed C chondrites (since at one time it was thought that their carbon content could be used to characterise them). There is, therefore, a wide range of oxidation levels shown by the chondrites, with a trend towards higher oxidation from top left to bottom right in the diagram.

The diagonal lines in figure 3.1 represent constant values for the bulk Fe/Si ratio. If all else were equal, meteorites lying on a given diagonal may have been made from the same source by differing levels of oxidation. Prior (1916) believed that this was true of all meteorites (§ 1.5.3), but figure 3.1 shows that four distinct Fe/Si ratios are required. Even the closely related H, L and LL chondrites cannot, therefore, have been made from the same material by a simple oxidation–reduction process. The only possibilities for such a transformation in the chondrite classes are that certain E chondrites could be converted into H and C chondrites by oxidation—or the reverse by reduction—but this must also be discounted when other element ratios are considered.

Other bulk major-element ratios (figures 3.2 and 3.3)

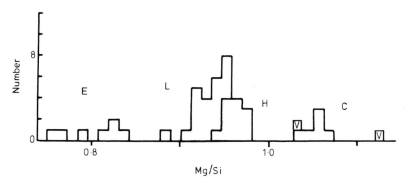

Figure 3.2. Histogram of Mg/Si values (by atoms) in chondritic meteorites. The values marked with a 'V' apply to CV chondrites. (Data from Von Michaelis *et al* 1969.)

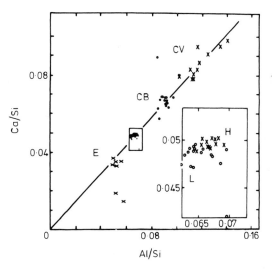

Figure 3.3. Plot of Ca/Si against Al/Si for chondritic meteorites. The full line indicates a Ca/Al value of 0·67. All ratios apply to numbers of atoms. (Data from Von Michaelis *et al* 1969, Van Schmus and Hayes 1974.)

60

support the same division (Ahrens 1964, 1965, Ahrens *et al* 1969). In addition, the C chondrites require further division on the strength of their Ca/Si and Al/Si values (Von Michaelis *et al* 1969, Van Schmus and Hayes 1974). The high Ca–Al group are termed the CV chondrites, and the remainder we will term the CB chondrites (following Van Schmus and Hayes 1974).

The important major-element ratios of the chondrite classes are summarised in table 3.2. (The metallic iron–silicon ratio is also given because this reflects the oxidation level.)

Table 3.2 Major-element atomic ratios in the chondritic classes (Van Schmus and Wood 1967, Van Schmus and Hayes 1974, Von Michaelis *et al* 1969).

Chondrites	Fe/Si	Fe^0/Si[†]	Mg/Si	Ca/Si	Al/Si
E	0·83	0·80	0·79	0·036	0·051
H	0·83	0·63	0·96	0·049	0·066
L	0·59	0·33	0·93	0·049	0·065
LL	0·53	0·08	0·94	0·050	0·065
CB[‡]	0·84		1·05	0·067	0·089
CV	0·75		1·07	0·084	0·115

[†] Fe^0 is metallic iron.
[‡] These are the C chondrites other than CV.

Oxidation–reduction processes cannot affect bulk element ratios. The possibility that certain E, H and C chondrites were made from the same material by oxidation–reduction processes can therefore be ruled out. In fact, there are no obvious processes by which any chondrite class could be converted into any other. Each seems to represent a unique set of formation conditions.

3.2.2. Mineralogy of the Chondrite Classes

Table 3.3 lists the main minerals found in E and ordinary chondrites (with estimates of their abundance), and C chondrites. The major minerals are olivine, pyroxene, feldspar, metal and sulphide. Most of the minerals are not pure compounds; they are solid solutions between end members of

61

Table 3.3. Principal minerals in chondritic meteorites (wt %)[†].

		E	H	L	LL
Olivine	$(Mg, Fe)_2SiO_4$	—	33–37	45–49	56–60
Pyroxene	$(Mg, Fe)SiO_3$	50–60	23–27	21–25	14–18
Diopside	$CaMgSi_2O_6$	—	4–5	4–5	4–5
Feldspar	$NaAlSi_3O_8$[‡]	5–10	9–10	9–10	9–10
Troilite	FeS	5–10	5–6	5–6	5–6
Kamacite	NiFe	15–25	15–17	6–8	1–2
Taenite	NiFe	trace	2–3	2–3	2–4

Additional important minerals in C chondrites

Serpentine	$Mg_3Si_2O_5(OH)_4$
Magnetite	Fe_3O_4
Spinel	$MgAl_2O_4$
Alumina	Al_2O_3
Melilite	$Ca_2MgSi_2O_7$

[†] The percentage abundance is given for equilibrated meteorites (types 5 and 6) based on thin-section measurements (Van Schmus 1969a). For the others this cannot be determined because the texture is too fine.
[‡] With some 10–15% $CaAl_2Si_2O_8$ and 1–6% $KAlSi_3O_8$.

definite composition. For instance, olivine $((Fe, Mg)_2SiO_4)$ is a solid solution of Fe_2SiO_4 (fayalite) and Mg_2SiO_4 (forsterite). The metal exists as two nickel–iron alloys, kamacite and taenite. Kamacite is an alloy with less than 7% Ni and is body-centred cubic (BCC) in structure. Taenite contains more than 25% Ni, and has a face-centred cubic (FCC) structure. Fine-grained mixtures of the two alloys, which are found in the centre of taenite grains and as small isolated grains in chondrules, are called plessite. The formation of these metal types is usually assumed to have occurred in a similar way to the analogous phases in iron meteorites (§ 3.5.2). In C chondrites, magnetite and serpentine (a complex hydrated silicate) are also important. Certain members of this class contain whitish aggregates of minerals rich in Ca and Al.

On the whole, the mineralogical differences between the chondrite classes, especially in the minor minerals, reflect their different oxidation levels. These have a considerable bearing on formation conditions, and we will briefly discuss each class in turn.

(i) *E Chondrites.* The most notable feature of the E chondrites is their high state of reduction. Not only is the iron in the metallic state, constituting 15–25% of the meteorite, but the dominant silicate is an iron-free pyroxene (enstatite). Cr, Mn and Ca exist as sulphides ($FeCr_2S_4$, daubreelite; MnS, alabandite; CaS, oldhamite), whereas in other meteorites they are present only as oxides or silicates. Similarly, sinoite (Si_2N_2O) and osbornite (TiN) are unique to the E chondrites. Even part of the silicon is reduced and is in solid solution in the metal; in this property, the E chondrites resemble two anomalous metal-rich meteorites (Horse Creek and Mount Egerton).

The E chondrites show a range of oxidation levels which parallel several chemical and petrological properties (Keil 1968, Larimer 1968). Those members with high Fe/Si ratios in figure 3.1 also have high sulphur abundances, and are more reduced than the low Fe/Si, low sulphur members. For example, E chondrites with high Fe/Si ratios have about 3·5% Si in their metal, compared with about 1% in the others. The high and low Fe/Si members are sometimes distinguished by the terms EI and EII (or EH and EL), respectively, but this is misleading in that it implies two separate groups, whereas they probably represent extremes in an unbroken sequence.

(ii) *H, L and LL Chondrites.* The ordinary chondrites are composed of both olivine and pyroxene, the relative amounts varying from class to class (table 3.3). Their higher oxidation level than E chondrites is apparent in that the silicates contain an appreciable proportion of iron end member, and the amount of this varies discretely between H, L and LL†. The amount of metal often distinguishes the three classes in the hand specimen. The LL chondrites also usually differ from the others in being more highly brecciated, that is, made of fragments of rock. There appears to be no Prior's law relationship

---

† On average, the olivine contains 19·3, 25·2 and 31·3% of the iron end member, and pyroxene contains 16·8, 20·9 and 25·2% of its iron end member for the H, L and LL groups, respectively (Van Schmus 1969a).

between the metal and silicates within each class, except possibly the LL chondrites (Keil and Fredriksson 1964, Sears and Axon 1976).

(iii) *CB Chondrites.* The important features of the mineralogy of the CB chondrites are their high oxidation state and their content of volatile substances; both of these properties display important trends within the group. All CB chondrites were formed in more oxidising conditions than the other chondrites, but there is some variation within the group. The volatile substances in question are carbon, sulphur and water. By geological standards H_2O itself is volatile, whilst carbon and sulphur form stable volatile compounds. Wiik (1956) defined three types. In weight per cent, and using slightly different nomenclature from that of Wiik, these types are as follows.

	C	H_2O	S
C1	3·54	20·08	6·20
C2	2·46	13·35	3·25
CO	0·46	0·99	2·37

In addition to being much poorer in volatiles, CO chondrites (named after Ornans) are slightly less oxidised than the others, as they contain metal and sulphide. In C1 chondrites the silicates, and other minerals, are hydrated—the major silicate is now serpentine $(Mg_3Si_2O_5(OH)_4)$—and this also applies to some extent to C2 chondrites. All of the iron in C1 chondrites, and most in C2 chondrites, exists as magnetite. The carbon is discussed in § 4.8. C1 chondrites tend to have a higher Fe/Si content than C2, but the major distinction is that C1 contain no chondrules. This may explain their apparently higher content of volatile substances, because chondrules are volatile-free.

The abundance of carbon, sulphur and H_2O and of many volatile trace elements, make CB chondrites particularly important. These substances are the first to be lost by any heating subsequent to their formation. Consequently, the CB

64

chondrites, and particularly the C1 chondrites, are usually considered to be the most primitive—in the sense of least altered—samples of the solar system known to man. We will discuss further grounds for this belief in § 4.1.

(iv) *CV Chondrites.* Except for their higher Ca/Si and Al/Si values, CV chondrites closely resemble CO chondrites in their composition and oxidation state. Van Schmus (1969b) pointed out that CV chondrites consist of large spongy chondrules in an abundant fine-grained matrix, whereas CO chondrites consist of closely packed small chondrules. In each case the matrix is essentially olivine, which is more Fe-rich than in ordinary chondrites. Several CV chondrites contain large (about 2 mm), whitish aggregates consisting of minerals which have a particularly high Ca and Al content, such as spinel, melilite and alumina.

3.2.3. Petrology of the Chondrite Classes

We have examined the mineralogical properties of chondrites as a guide to oxidation–reduction states, because these have a bearing on the formation process. Petrological studies deal with the physical structures within meteorites. They concern not only originally formed structures, but also equally significant structures which formed later in the history of the meteorite.

   Under the microscope in thin section, most chondrites are seen to consist of large (say 2 mm), circular chondrules, with intricate internal structures. Irregular opaque grains of metal and sulphide usually occur outside chondrules and opaques are set in a very fine silicate matrix (figure 3.4). Superimposed on this original texture are two kinds of alteration usually ascribed to shock impact and metamorphism (that is, slow crystallisation at elevated temperatures). The three aspects of the petrology of chondrites which will be examined here, therefore, are (i) the chondrules, (ii) metamorphism, and (iii) thermal and mechanical alteration resulting from shock.

(i) *Chondrules.* The abundance of chondrules in the major classes of meteorites makes them useful recognition features, since they have rarely, if ever, been observed in terrestrial

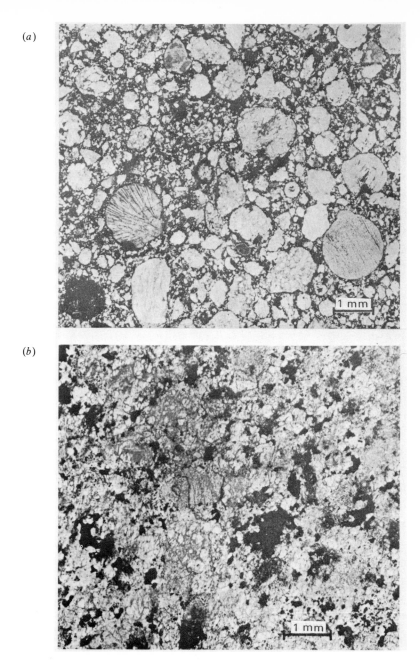

(a)

(b)

Figure 3.4.(*a*) Thin section of the Bishunpur (L3) chondrite, showing the representative texture of type 3 chondrites. Note the distinctive chondrules, which range in size from about 0·2 mm to just over 1 mm, including the eccentro-radiating pyroxene chondrule left of centre, and the numerous chondrule fragments (Van Schmus and Wood 1967). (*b*) Thin section of the Peace River (L6) chondrite showing the texture of type 6 chondrites. Note that chondrules have merged with the matrix, and are therefore difficult to identify, and that the matrix has a coarser texture (Van Schmus and Wood 1967).

rocks. Figures 1.4 and 3.4 show some examples of the various textures observed. All are spherical in shape and some contain glass, although this may have devitrified into fine crystals. The textures inside chondrules suggest that they were liquids which were suddenly quenched: they consist of thin fibres, grains or lathes which are imperfectly grown and often have cavities containing glass. The bulk composition of over 500 individual chondrules has been measured by Osborne *et al* (1973, 1974) and Walter (1969). They show correlations between Ir, Al and Sc, which have been interpreted as fractionations due to condensation or evaporation during melting. Most workers favour a melt–quench mechanism for the origin of chondrules, but there is debate as to when, and how, the melting occurred. Lightning discharges (Whipple 1966), supercooling (Blander and Katz 1967) in the primordial nebula, and impact melting (Urey 1961, Cameron 1973, Dodd 1976) are strong possibilities. Similar structures (King *et al* 1972) which have been shocked and mixed by meteorite impact have been found in lunar samples, and this provides circumstantial evidence for impact melting. The chondrules in meteorites are, however, better sorted and more abundant than in lunar rocks; hence they accreted differently (Dodd 1976).

(ii) *Metamorphism.* Metamorphism means the slow crystallisation, or crystal growth, at elevated temperatures. The signs of metamorphism in chondrites have been recognised for a century (Sorby 1877), although there is still some doubt as to whether it occurred as part of the formation process or subsequent to it (Dodd 1969b, King *et al* 1972). The textural effects are a coarsening of the matrix material by crystal growth, and an obscuring of the boundary between chondrules and the matrix (figure 3.4(*a*), cf figure 3.4(*b*)). The mineralogical trends are mainly homogenisation of the mineral composition: the few little-affected meteorites show wide variations in their mineral composition, while the majority are noted for their uniformity in this respect. The effect of metamorphism is therefore to simplify and to bring the assemblage nearer to equilibrium (Dodd 1969b). In addition, there may be compositional changes, since metamorphosed

67

meteorites are usually depleted in volatile substances, sometimes dramatically so.

These trends have been incorporated into a descriptive scheme by Van Schmus and Wood (1967). The least altered of all meteorites, because they contain abundant volatile substances, are the CB chondrites and, as we saw above, Wiik (1956) sorted these into smaller groups on this basis. For the CV and CB chondrites, Van Schmus and Wood (1967) have defined petrological types 1, 2, 3, and 4; and a type 5 has recently been added by Binns *et al* (1977). A few ordinary chondrites contain as much carbon as the CV and CB chondrites of types 3 and 4, and there appears to be some overlap in the extent of metamorphism they have suffered. These ordinary chondrites are also unequilibrated in terms of their highly inhomogeneous silicates and sharp chondrule outline. They are assigned to petrological types 3 and 4. Most ordinary chondrites are more highly metamorphosed, and are assigned to petrological types 5 and 6. In addition, one partially melted L chondrite (Shaw) has been assigned to petrological type 7 by Dodd *et al* (1975). EI chondrites are of petrological type 4, and EII chondrites are of type 6; the intermediate St Marks and Saint Sauveur have been assigned to type 5. There may also be a E7 chondrite (Olsen *et al* 1977). The definitions of the petrological types are summarised in table 3.4. Following Van Schmus and Wood (1967), the petrological type of a meteorite is usually given with its class (for example, L5, H6, E4 and CV3, etc).

The question naturally arises as to whether the higher petrological types can be produced by metamorphism of the lower types, or whether the types represent varying degrees of metamorphism of different starting materials. For the E and C chondrites the latter is clearly the case. As we saw earlier (figure 3.1), E and C chondrites show a range of Fe/Si values; EI (E3) have higher values than EII (E6), and, similarly, C1 have higher values than C2. Experience with terrestrial rocks shows that metamorphism seldom, if ever, changes the ratios of non-volatile major elements (Urey 1961). It seems impossible, therefore, that in these classes the higher petrological types can be made from the lower types. Dodd (1976) has suggested that this also applies to ordinary chondrites.

Table 3.4. Definition of the petrological types in chondrites (Van Schmus and Wood 1967)†.

	1	2	3	4	5	6
Silicate homogeneity		5% mean deviation		5% mean deviation	Uniform	
Pyroxene structure		Mainly monoclinic			Orthorhombic	
Feldspar		Absent			Microcrystalline	Grains
Glass		Clear isotropic		Turbid	Absent	
Chondrule texture	None	Very sharp		Well defined	Readily visible	Poorly defined
Matrix texture	Fine and opaque	Opaque		Trans. Microcryst	Recrystallised	
C content	2·8%	0·6–2·8%	0·2–1·0%	0·2%		
H$_2$O content	20%	4–18%	2%			

† In recent years, meteorites which show signs of having started to melt have been ascribed to type 7.

However, the variation in the Fe/Si ratio within the classes is very much less, and it has yet to be established that it is significant.

The best estimates of the temperatures experienced during metamorphism are probably those calculated from oxygen isotope measurements (§ 5.4.2). These suggest that the maximum temperature reached by ordinary chondrites during metamorphism was 960 K for type 4, 1120–1320 K for types 5 and 6, and 1570 K for Shaw (L7). Most of the estimates based on mineral chemistry of the ordinary chondrites have been challenged (for example, see Wasson 1972 for a review), but for the E chondrites estimates range from about 970 K for type 4 to about 1120 K for type 6 meteorites (Larimer and Buseck 1974).

Estimates of the rate at which the meteorite cooled after metamorphism are important because of their bearing on the burial depth in the meteorite parent body: higher cooling rates suggest shallower burial depths. Such estimates have been made from the compositional zoning in the taenite—as for iron meteorites (§ 3.5.2)—and from ^{244}Pu tracks†. The metallographic cooling rates were $1-10\,°C\,(10^6\,yr)^{-1}$ for petrological types 5 and 6, and $0 \cdot 1-1 \cdot 0\,°C\,(10^6\,yr)^{-1}$ for petrological type 3 (Wood 1967). The ^{244}Pu track results are given in table 3.5.

Table 3.5. Cooling rate estimates based on ^{244}Pu tracks (Pellas and Storzer 1976).

Meteorite	Class	Cooling rate between 300 and 550 K $(K(10^6\,yr)^{-1})$
Beaver Creek	H4	1·6
Kiffa	H4	1·6
Sena	H5	1·5
Allegan	H5	1·4
Estacado	H6	1·4
Kernouve	H6	1·4
Guarena	H6	1·4
Peetz	L6	0·9
Tillaberi	L6	0·8
Shaw	L7	0·8
St Severin	LL6	0·9
St Mesmin (xenolith)	LL7	0·7

The results of the two methods agree reasonably well, except that they show the reverse trend with petrological type. The track data, however, are subjected to fewer uncertainties and are probably superior.

(iii) *Thermal and Mechanical Alteration.* The mechanical alterations are usually caused by shock from collisions in space. Many chondrites contain black veins. These are thin, black lines which may cross the meteorites as a single vein, branch in a number of directions, or even soak through whole regions in the stone. They are similar to features in terrestrial

† The principles of the method are that (i) track densities are proportional to age, and (ii) different minerals start track retention at different temperatures. The temperature against time curve can then be used to derive the cooling rates. Only grains adjacent to phosphates (in which the ^{244}Pu was located) can be used.

rocks surrounding faults, and in specimens which have been artificially shocked. Some meteorites are entirely blackened by the process. Under the microscope the vein is seen as an opaque glass, sometimes with very fine microcrystals of silicate, and numerous small sulphide globules. The metal is particularly susceptible to shock pressures and the associated increases in temperature. Metallic structures can be used as a sensitive indication of the levels of shock experienced, and these can be related directly to the presence of veins and the darkness of the stone (Heymann 1967).

3.3. Achondrites

3.3.1. The Achondrite Classes

The main distinction between achondrites and chondrites is that the achondrites have different abundances of Ca and similar elements, almost no metal or sulphide and, by and large, are composed of crystals grown from melts. When the amount of CaO (or the Ca/Si ratio) in achondrites is plotted against $FeO/(FeO+MgO)$—a readily measured indication of the proportion of Fe in the silicates—the achondrites divide themselves into five groups with appreciable gaps between them (figure 3.5). The two Ca-rich classes with more than 5%

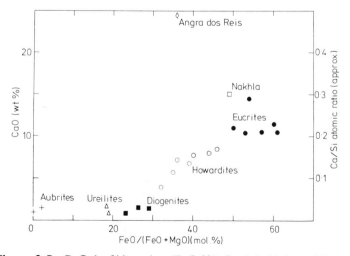

Figure 3.5. CaO (wt%) against $FeO/(FeO+MgO)$ (mol.%) for achondrites (after Mason 1967b).

CaO (or where the Ca/Si ratio is greater than 0·1) are the eucrites and the howardites (named after E C Howard). There are three Ca-poor classes: the diogenites (named after a Greek philosopher who suggested that meteorites were extra-terrestrial); the ureilites (after the Nova Urei meteorite); and the aubrites (after the Aubres meteorite). A number of anomalous achondrites are listed, with references, in table 3.1.

3.3.2. Ca-Rich Achondrites

As a consequence of their composition, the Ca-rich achon-drites have a markedly different mineralogy from the chon-drites (Mason 1967, Duke and Silver 1967). Calcium resides mainly in the plagioclase (Ca, Na feldspar): roughly half the meteorite consists of this mineral, the remainder being pyroxene. A more detailed breakdown of the mineralogy is

Table 3.6. Mineral compositions of the Ca-rich achondrites and mesosi-derites (%) (Duke and Silver 1967, Mason and Jarosewich 1973).

		Eucrite (Sioux County)	Howardite (Yurtuk)	Mesosiderites (Vaca Muerta†)	(Barea‡)
Plagioclase solid solution					
Albite	$NaAlSi_3O_8$	3·6	—	—	1·0
Anorthite	$CaAl_2Si_2O_8$	32·8	25·4	29·2	10·4
Pyroxene solid solution					
Enstatite	$MgSiO_3$	17·6	35·4	34·9	
Ferrosilite	$FeSiO_3$	33·3	24·5	24·7	26·6
Wollastonite	$CaSiO_3$	7·7	—	—	
Accessory minerals					
Quartz	SiO_2	2·6	—	1·9	2·6
Olivine	$(Mg, Fe)_2SiO_4$	—	9·9	—	—

† Silicate portion recalculated to 100% to allow comparison with the Ca-rich achondrites.
‡ 56·6% metal, 1·6% troilite and 2·0% other accessory minerals.

given in table 3.6. They contain very little metal and are much more coarsely crystalline than the chondrites. Rose's (1825) conclusion, that these meteorites were igneous in origin, has never been challenged. In terms of composition, mineralogy and texture, they resemble the igneous rocks which share the eucrites' name, although terrestrial eucrites are higher in sodium and lower in iron. All howardites, and most eucrites, are brecciated, and fragments of eucrites occur in howardites (Wahl 1952).

3.3.3. Ca-Poor Achondrites

Except for their almost total lack of Fe and S, the aubrites resemble the E chondrites in many respects. They are essentially pure enstatite, with very little metal, troilite, schreibersite $((Fe, Ni)_3P)$ and oldhamite (CaS). As in the E chondrites, the Ca-poor achondrites indicate a highly reducing environment. Eight of the nine aubrites are brecciated.

The diogenites are almost pure pyroxene (22–25% Fe end member) with very minor amounts of other minerals present in ordinary chondrites: olivine troilite, metal, and chromite. They are also almost all brecciated (Mason 1962) and recently one was discovered which contained a eucrite fragment.

The ureilites are noted for their carbon content (1% in Nova Urei), much of which is in the form of diamonds, to which Lipschutz (1964) has ascribed a shock origin from impact in space. They consist of grains of olivine (21% Fe end member) and Ca-bearing pyroxene sitting in a network of carbonaceous veins. These contain metal and troilite, but in smaller quantities than the ordinary chondrites (Vdovykin 1970, Haverö 1972, Kenna 1976).

3.4. Stony-Iron Meteorites

3.4.1. The Stony-Iron Classes

Stony-iron meteorites are traditionally defined as having approximately equal proportions of stony material and iron. At various times, four groups of stony-iron meteorites have been defined, but since two of these contain only one meteorite each we need here consider only two in any detail: the pallasites and the mesosiderites. They differ in many ways; their similar proportions of stone and metal being their only common property. The pallasites are so called because the first of their kind to be widely known was the Pallas iron (figure 1.1). Mesosiderites derive their name from their intermediate position between stones and irons ('siderites').

3.4.2. Pallasites

The most striking feature of the pallasites is that their metal forms a continuous framework in which are set crystals of the silicate olivine (figure 3.6). The metal is composed of zoned

(*a*)

(*b*)

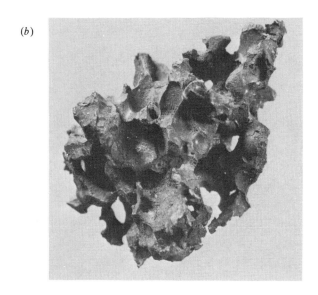

Figure 3.6.(*a*) The Thiel Mountains pallasite. A network of metal (light kamacite and darker taenite) encloses nodules of olivine which, in this case, are globular. The specimen is about 7 cm in its longest dimension. (Smithsonian Institution photograph). (*b*) Specimen of the Krasnojarsk pallasite (the Pallas iron shown in figure 1.1) at Leicester University. The silicates have been removed, probably by terrestrial erosion, to leave a single metallic network. The specimen is about 5 cm in its longest dimension. (Photograph courtesy of Dr A A Mills.)

taenite, usually with a poorly developed Widmanstätten pattern. Pallasites contain much troilite, and some schreibersite and phosphate. The silicate is 0·5–1 cm olivine nodules, which may be angular (such as Eagle station), or globular (such as Thiel Mountains; figure 3.6). Most have olivine with 11–12 mol.% Fe end member and metal, with 8–13% nickel. Springwater, Eagle Station and Itzawisis, however, vary with regard to these properties, so Yavnel (1958) has proposed that they should be considered as a subgroup of the pallasites. Oxygen isotope studies confirm that Eagle Station and Itzawisis formed in an environment different from those in which the other pallasites formed.

Mason (1963b) has proposed that the apparent relationship between the olivine and metal compositions suggests a common origin for the two phases, although others have suggested that the metal was injected as a liquid from a separate source. In fact, most authors favour an igneous origin for this class of meteorite. The biggest problem is why the silicates did not float up and out of the metal. It has been suggested that the liquid began to solidify immediately after intrusion (Scott 1977). Using the methods described in § 3.5.2, Buseck and Goldstein (1969) estimated a cooling rate of $1\,°C\,(10^6\,yr)^{-1}$ for the pallasites, which is lower than for most iron meteorites and suggests greater burial depths.

3.4.3. Mesosiderites

In mesosiderites the metal is not continuous, but is present as individual grains which constitute half of the meteorite (figure 3.7). The taenite is nickel-rich (more than 35%), and has therefore not decomposed into plessite. The metal shows no sign of a Widmanstätten pattern: where kamacite has precipitated its orientation is random and governed by grain boundaries, etc (Powell 1969). The grains show a composition–dimension relationship similar to the zoned taenite in chondrites, which, if due to the same mechanism, suggests a cooling rate of $0·1\,°C\,(10^6\,yr)^{-1}$, which is very much lower than that for iron meteorites.

The silicates are pyroxene and plagioclase, with minor amounts of olivine. Tridymite (SiO_2) is present in most

Figure 3.7. The Mincy mesosiderite. The metal and silicates are in approximately equal proportions, but the metal (light) does not enclose the silicates. The specimen is 20 cm across. (American Museum of Natural History.)

mesosiderites, which shows that these meteorites have never experienced pressures greater than 3 kbar (Mason 1962). The mesosiderites are noted for the extreme heterogeneity of their silicate composition. This suggests rapid initial cooling, contrary to the cooling rate indicated by the metal, but the metal structures may have been determined by subsequent metamorphism which did not affect the silicates (Powell 1969). The mineral composition of the Barea mesosiderite is given in table 3.6.

Chemically and mineralogically, the silicates of the mesosiderites are very similar to those of the howardites (table 3.7), which led Prior (1918) to suggest that the former were

Table 3.7. Ni, Ga and Ge contents of the major iron meteorite groups (after Scott and Wasson 1975).

Group	Ni(%)	Ga (ppm)	Ge (ppm)
IA†	6·4–8·7	55–100	190–520
IB	8·7–25	11–55	25–190
IIAB	5·3–6·4	46–62	107–185
IIIAB	7·1–10·5	16–23	27–47
IVA	7·4–9·4	1·6–2·4	0·09–0·14
IVB	16–18	0·17–0·27	0·03–0·07

† Separate figures are given for IA and IB because, although these are members of the single IAB group, IB is best considered as a high-Ni, low-Ga, Ge tail on the main IA group. There are 82 IA members and eight IB members.

mechanical mixtures of howardites and metal. However, the Ca/Al ratio of stony meteorites, including howardites, is remarkably constant (1·09) and significantly higher than for the mesosiderites (0·88), so that these two classes may have had a similar evolution from different starting materials (Mason and Jarosewich 1973, Powell 1971, MaCarthy and Ahrens 1971).

3.5. Iron Meteorites

3.5.1. The Iron Meteorite Classes

Within the limits prescribed in § 3.1, the most successful classification scheme for iron meteorites is that based on gallium, germanium and nickel content (Goldberg *et al* 1951, Lovering *et al* 1957, Scott and Wasson 1975). The important feature of the distribution of Ga and Ge is that, instead of being present over an entire range of concentrations, they show 'preferred' or 'quantised' values. This is illustrated in figure 3.8. The major iron meteorite groups in order of decreasing Ga and Ge abundance are IAB, IIAB, IIIAB, IVA and IVB†. To some extent Ni also shows quantisation; IIAB

† IVA and IVB are in no way related. Occasionally the terms IA and IB, IIA and IIB, IIIA and IIIB are used to describe Ni-poor and Ni-rich members of IAB, IIAB and IIIAB, respectively.

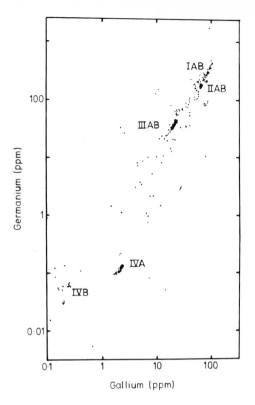

Figure 3.8. Plot of germanium abundance against gallium abundance in iron meteorites. All meteorites are plotted but only the major groups are indicated. (Data from Wasson 1974.)

irons have a lower Ni content (5·5%), and IVB higher (16–18%), than IIIAB and IVA irons which are· very similar (7–10% Ni). A number of minor groups have also been defined, with structure and mineralogy sometimes being incorporated into the class definition (table 3.8).

3.5.2. *Iron Meteorite Structures*

Before accurate trace element analysis made the classification scheme outlined above feasible, the most widely accepted classification was based on structure. The Widmanstätten pattern, the discovery of which was outlined in Chapter 1, consists essentially of plates of kamacite in an octahedral orientation with the interstices filled with taenite (figure 3.9). Kamacite and taenite were defined in § 3.2.2; they are known

Table 3.8. Mineral abundances in the major iron meteorite groups† (after Scott and Wasson 1975).

Group	Troilite (FeS)	Schreibersite ((Fe, Ni)₃P)	Cohenite ((Fe, Ni)₃C)	Graphite (C)
IAB	2N	2, M, rh	3, M	2 (M)
IIAB	1 (N)	2→3, rh	1	dc
IIIAB	1→3, N	1→3, rh, M	0	0→1
IVA	2, (N)	0→1	0	0
IVB	1	1→2	0	0

† Key: N, nodules; M, macroprecipitates; rh, rhabdites; dc, carbon which has formed by carbide decomposition. 0, absent; 1, sparse; 2, common; 3, ubiquitous. The arrows indicate systematic trends with Ni content.

to metallurgists as α and γ nickel–iron, respectively. The coarseness of the Widmanstätten pattern is expressed as the thickness of the kamacite plates (its 'band width'), and meteorites may accordingly be described as coarsest, coarse, medium, fine and finest octahedrites. Where the specimen is too coarse for this structure to show up (the whole meteorite is effectively from a single kamacite crystal) it is termed a hexahedrite. In these specimens, the most characteristic features visible on the polished and etched section are Neumann bands, which are oriented along the faces of a cube (hexahedron). At the other extreme, where the meteorite is mainly taenite, the hand specimen may appear featureless, and the specimen is then termed an ataxite (Greek, 'without structure'). Intermediate between ataxites and the finest octahedrites are 'plessitic octahedrites'.

The classification of irons based on structure was gradational, there being few, if any, hiatuses in the band width range. Consequently, there has been much disagreement over the best definition of each structural class. The definition used here is due to Buchwald (1975), in which the band width intervals are equal on a logarithmic scale. In this scheme, hexahedrites have band widths in excess of 50 mm; coarsest octahedrites 50–3·3 mm; coarse octahedrites 3·3–1·3 mm; medium octahedrites 1·3–0·5 mm; fine octahedrites 0·5–0·2 mm; and finest octahedrites, less than 0·2 mm. When band widths are defined in this way, there is a very good

79

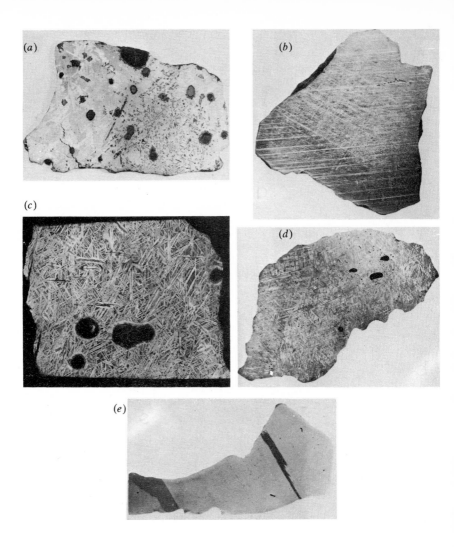

Figure 3.9. Iron meteorite structures revealed by etching polished
sections. (*a*) The Rifle IAB iron meteorite containing 7·2% Ni with a
coarse octahedrite Widmanstätten pattern. It contains many inclusions,
such as sulphide, carbide and graphite. In the right half of the specimen,
carbides lie along the centre of the Widmanstätten plates, emphasising the
structure; its longest dimension is 6 cm. (*b*) The Bennett County IIAB iron
meteorite with hexahedrite structure (5·3% Ni). The fine lines crossing the
specimen are Neumann bands. A few large phosphide grains can also be
seen (black). Longest dimension is about 20 cm. (*c*) The Mount Edith
IIIAB iron with medium octahedrite structure (9·4% Ni). The large globular inclusions are sulphide. A number of very thin sulphide and phosphide needles are also present. Longest dimension is 18 cm. (*d*) The
Gibeon IVA iron meteorite with fine octahedrite structure (7·7% Ni). The

correlation between structure and class. IA are mainly coarsest octahedrites, IIA are hexahedrites, IIB are coarsest octahedrites, IIIAB are medium octahedrites, IVA are fine octahedrites, and IVB are ataxites (figure 3.9). In addition, IIC are all plessitic octahedrites, and many meteorites which are classed as being chemically anomalous also have anomalous structures.

The formation of the Widmanstätten pattern is best understood by reference to the low-temperature part of the nickel–iron phase diagram (figure 3.10). Consider an alloy of

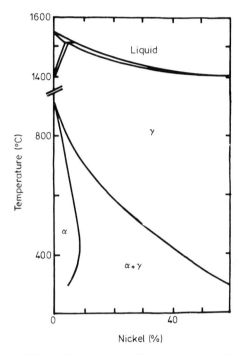

Figure 3.10. Nickel–iron phase diagram. α is a body-centred cubic alloy also known as kamacite, and γ is a face-centred cubic alloy known as taenite.

---

dark inclusions are sulphide. Longest dimension is 70 cm. (e) The Iquiqui IVB iron meteorite. The specimen is an ataxite and is structureless (16% Ni). The darker areas are not different phases, but different sheens in a macroscopically homogeneous alloy. They are common in IVB irons (R S Clarke 1977 personal communication) and probably of shock origin. Longest dimension is 9 cm. (All photographs from the Smithsonian Institution.)

8% nickel. Above approximately 850 °C, this exists entirely as taenite. On cooling, it enters the $\alpha + \gamma$ field and α begins to precipitate. The precipitation of α requires both the diffusion of Ni to produce low-Ni kamacite, and rearrangement of the lattice from FCC to BCC. Minimum rearrangement will be required if the kamacite is oriented with one of its crystal faces parallel to the octahedral face of the parent taenite. As the temperature drops, the amount of kamacite increases and, since the overall nickel content is constant, the Ni content of the taenite increases. The kamacite grows into plates with an octahedral configuration, and residual parent taenite is left trapped in the interstices between the plates. The process stops when diffusion no longer takes place (at about 350 °C) and the resulting structure is that of the octahedrites. An alloy of about 6% Ni cools through the two-phase region and into the α region, so that by the time diffusion ceases, the meteorite consists entirely of kamacite, and a hexahedrite is produced. At the other extreme, with about 20% Ni, diffusion is already very sluggish when the two-phase region is entered and only a highly localised diffusion takes place. In these instances, the ataxite structures result. Diffusion does not cease abruptly, but can take place over smaller distances as the temperature drops. Nickel concentration gradients therefore build up in the α and, especially, the γ regions. The presence of the Widmanstätten pattern therefore depends on the bulk Ni content of the iron meteorite. Its coarseness is also governed by the rate at which the meteorite cooled: coarser textures require slower cooling rates.

The formation of the Widmanstätten pattern may be modelled numerically (Wood 1964, Goldstein and Ogilvie 1965). The diffusion process is described by Fick's second law:

$$\frac{\partial c}{\partial t} = \frac{\partial}{\partial x}\left(D\frac{\partial c}{\partial x}\right). \tag{3.1}$$

where c is the concentration at a distance x after time t. The diffusion constant, D, is also a function of Ni concentration, being much higher in taenite than in kamacite. A relationship between temperature, T, and time is also required. It is simplest to assume exponential cooling, that is, $T = T_0 \exp(-\tau t)$. The numerical solution of these equations enables the cooling rate to be estimated. The results may be handled in a number of ways. In figure 3.11 the band width has been calculated as a function of bulk nickel content and the resulting curves are compared with the data for 67 iron meteorites.

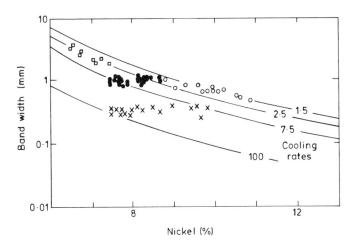

Figure 3.11. Band width against nickel data for 67 iron meteorites. Open squares represent group IAB; full circles, group IIIA; open circles, group IIIB, and crosses, group IVA. Theoretical band width against nickel curves for various cooling rates are superimposed, in units of $°C \, (10^6 \, yr)^{-1}$ (Goldstein and Short 1967a, b).

Two intra-group trends are apparent: most members of group IAB cooled at a similar rate of $2 \cdot 5 \, °C(10^6 \, yr)^{-1}$, whereas in group IVA a wide spread of cooling rates, between 100 and $7 \cdot 5 \, °C(10^6 \, yr)^{-1}$, is observed. IIIAB shows mixed behaviour; since IIIA cooled over a range of cooling rates ($7 \cdot 5–1 \cdot 5 \, °C$ $(10^6 \, yr)^{-1}$), but members of the IIIA end of the IIIAB group cooled at an essentially uniform rate of $1 \cdot 5 \, °C(10^6 \, yr)^{-1}$. The method is not applicable to classes with no Widmanstätten pattern (IIAB or IVB), or to meteorites where the kamacite plates have grown so large that their growth has become limited by impingement of adjacent plates (that is, meteorites with less than $8 \cdot 2\%$ Ni). However, the cooling rates of four

IIAB irons have been determined by theoretical modelling of the nickel profiles around rhabdites (small (Fe, Ni)$_3$P crystals). Uwet and Lombard were found to have cooling rates of $0.5\,°C(10^6\,\text{yr})^{-1}$, and Tocopilla and Coahuila $5\,°C(10^6\,\text{yr})^{-1}$; so this group resembles IVA in showing a variation by a factor of ten in the cooling rates (Randich and Goldstein 1975).

Willis and Wasson (1977) claim that the range of cooling rates observed in IVA and, by implication, IIIA, may be artifacts due to uncertainties in the diffusion coefficient of Ni and the phase diagram. This is contrary to the conclusions of Moren and Goldstein (1977) who use their own diffusion coefficients and phase diagrams and find a variation by a factor of ten in the cooling rate estimates, regardless of the method used.

Cooling rates may be used to determine the depth at which an iron meteorite was buried as it cooled. The equation for heat conduction in a spherically symmetrical body is

$$\rho C_\text{p} \frac{\partial T}{\partial t} = \frac{1}{r^2} \frac{\partial}{\partial r} \left(r^2 K \frac{\partial T}{\partial r} \right) + \frac{\partial Q}{\partial t}, \qquad (3.2)$$

where ρ is the density, C_p is the specific heat, K is the thermal conductivity of the parent body (which is usually taken to be chondritic), and T is the temperature at depth r after time t. Q is the heat available. Wood (1967) and Fricker *et al* (1970) have presented the results of numerical solution of this equation. Wood's (1967) results show that for bodies less than about 400 km in diameter, heating from long half-life radioactive isotopes makes little contribution to the internal temperature. The metallographic cooling rates for stony meteorites suggest that, when they cooled, they were buried 20–150 km in a body greater than 90 km in diameter. The charged-particle track studies suggest that the bodies were greater than 120–200 km in size. The iron meteorites indicate bodies of at least 120 km for IAB, and require that IIAB, IIIAB and IVA were radially dispersed throughout 200, 150 and 100 km bodies. These values assume that the temperatures of the bodies were initially above 600 °C, and that the heat source was removed abruptly. They are therefore somewhat speculative.

3.5.3. Mineralogy

The important accessory minerals in iron meteorites are sulphides, phosphides, carbides and carbon. For further comments on these minor minerals, the reader may be referred to Mason (1972). The major sulphide is troilite (FeS), which occurs as nodules or lamellae ('Riechenbach lamellae'). The nodules are usually macroscopic, perhaps 5 mm across, and sometimes form complex assemblages with other sulphides or graphite (figure 3.9). Their sulphur content is too high for them to be ex-solution products, and they were possibly formed in the residual melt after the metal began to solidify.

The phosphide $(Fe, Ni)_3P$ (schreibersite) occurs in two major forms: as globular masses on grain boundaries, and as microscopic rhombohedral crystals called rhabdite. It also occurs as long, thin needles ('Brezina lamellae'). The morphology is due to different nucleation mechanisms (Reed 1965, Doan and Goldstein 1969).

Carbon exists (i) as large (centimetre-sized) graphite nodules (figure 3.9) which are usually complex assemblies with sulphides and silicates; (ii) as cliftonite, in the form of small graphite squares resulting from the decomposition of carbides (Brett and Higgins 1969); (iii) as diamond, as found in Canyon Diablo, which was discussed earlier. Cohenite $((Fe, Ni)_3C)$ resembles schreibersite in shape, colour and texture, but is not restricted to grain boundaries and, when present, tends to be larger (figure 3.9). The Fe–Ni–C phase diagram indicates that cohenite is unstable. This led Ringwood and others (for example, Ringwood 1960b) to suggest that the high pressures inside very large (Moon-sized) meteorite parent bodies stabilised the cohenite. However, Brett (1966) has shown that its presence is actually due to its slow decomposition.

An attempt to relate mineralogy to meteorite class is shown in table 3.8. Of the five major groups, only IAB contains appreciable amounts of carbon minerals. Except in IAB, schreibersite is invariably more abundant in nickel-rich members of the class, and only in these does it occur in macroscopic forms. Troilite is similarly most abundant in the nickel-rich members of IIIB. It is particularly scarce in IVB (see also § 5.2).

Only two classes of iron meteorite contain members with an appreciable silicate content: IAB and IIE. In IAB, the silicates, which are usually around 5% by volume, resemble H chondrites in their bulk composition and mineralogy, but the silicates have lower iron contents (Scott and Wasson 1975, Bunch *et al* 1970). Four of the eleven IIE irons contain 5–10 vol.% silicate. Usually they are rounded and richer in iron than are IAB. Netschaëvo silicates have a chondritic bulk composition and even contain chondrules. The presence of silicates in iron meteorites is frequently used as an argument against complete melting of the metal.

3.5.4. *Thermal and Mechanical Alteration*

Thermal and mechanical alteration are the result of comparatively recent events. Often they are associated with dramatic events; for example, crater-forming meteorite impact with the Earth. Specimens of meteorites found near craters often show disrupted structures and lines of shear which can be traced over many centimetres. Deformation is scarce in meteorites not associated with craters (Buchwald 1975, Perry 1944, § 5.1.1). This is not so with the symptoms of less violent shock. Neumann bands are the result of shocks where the maximum pressure was less than 93 kbar, such as that which would result from collisions between metre-sized objects in space. They appear in kamacite wherever it occurs. More intense shock (greater than 130 kbar) changes the appearance of the kamacite completely, giving it a characteristic 'woody' appearance. Virtually all IIIAB irons and many IVA irons have this shock transformation structure. Since it is so common in one or two groups, and yet scarce in others, it seems most probable that the shock responsible was associated with the break-up of meteorite parent bodies.

Thermal alteration is also common in iron meteorites, and it occurs with a considerable range in intensity (Buchwald 1975). Often it is associated with the shock symptoms discribed above. Heating of the surface during passage through the atmosphere actually melts some of the material, and it destroys the structure within about a centimetre of the surface (§ 2.2.3). When the meteorite as a whole has been heated, the event has obviously been extraterrestrial. A number of IIA irons which have been reheated (about 25 out of 39), were once known as nickel-poor ataxites, because their hexahedral structure had been completely destroyed. Many IVA irons have undergone similar heating. The effects of reheating are not always readily recognised, especially when combined with the effects of shock, and laboratory experiments are needed to identify them (Brentnall and Axon 1962, Jain and Lipschutz 1968).

4. Compositional Properties

4.1. Introduction

The bulk of any meteorite is made up of a few elements, such as Si, Mg, Fe and O. Their abundance in representatives of each class of stony meteorite is given in table 4.1, together with the composition of the solar atmosphere. Since the units are different we have normalised the values to $Si = 20$ to enable comparisons to be made. The first feature to note is that the chondrites, whose analyses are given on the left of table 4.1, closely resemble the Sun. This extends to most elements present at the parts per million (ppm) level (trace elements). Rigorous comparison is best made by plotting one element against another. For C1 chondrites—the least altered class of meteorites (§ 3.2.2)—the agreement could hardly be better (figure 4.1). In contrast, the achondrites, on the right of table 4.1, are quite different in composition (note particularly Mg, Ca and Al). The chondrites are the only materials we have which show such agreement. For example, a cursory examination of the abundance of a few major elements in lunar rocks, or terrestrial rocks, show very different patterns. Lunar rocks are enriched in Ca, Al and Ti and depleted in Fe and Mg. Such a difference may be explained if the Moon had once been molten, so that the low-density, Ca–Al-rich minerals floated on top of the denser Fe–Mg-rich minerals. Similarly, the elements which concentrated in the metal, and descended to the core of the Earth and are depleted in the crust. On the other hand, those elements which concentrate in low-density surface rocks are over-abundant in the Earth's crust. Although the mechanisms outlined here are not the only ones possible, it seems clear that the major differences between the rocks in question and the Sun indicate some form of alteration on a large scale. The chondrites therefore appear

88

Table 4.1. Compositions of representatives of the stony meteorite classes (in wt%) and the Sun (in arbitrary units, with Si = 20).

| | Chondrites | | | | | | | | | Sun | Ca-rich achondrites | | | Ca-poor achondrites | |
	EI Indarch	EII Pillister	H Guarena	L Leedey	LL Cherokee Springs	CV Mokoia	CO3 Ornans	CO2 Mighei	C1 Orgueil		Eucrite Luotolax	Howardite Bununu	Diogenite Johnstown	Urelite Nova Urei	Aubrite Norton County
Si	16·47	10·50	17·18	18·86	19·08	15·62	15·60	13·00	10·55	20·00	23·54	22·76	25·00	18·57	25·40
Mg	10·54	12·62	14·15	15·03	15·25	14·40	14·66	11·73	9·54	20·95	9·90	8·57	15·60	22·23	24·55
Fe	33·15	22·05	24·24	18·01	15·88	19·62	25·83	21·20	18·28	16·64	13·20	13·42	12·91	15·64	1·60
Al	0·77	1·14	1·08	1·16	1·12	1·32	1·35	1·14	0·87	1·70	4·78	4·69	0·76	0·23	0·32
Ca	0·89	0·94	1·12	1·30	1·32	1·83	1·40	1·18	0·87	1·44	4·88	4·84	1·00	0·57	0·69
Na	0·75	0·067	0·067	0·08	0·08	0·03	0·41	0·037	0·05	1·21	0·28	0·25	0·03	0·13	0·09
Ni	1·89	1·68	1·74	1·21	0·97	—	1·36	—	—	0·96	—	0·06	0·05	0·12	0·04
Cr	0·23	0·29	0·38	0·36	0·35	0·36	0·38	0·25	0·25	0·25	0·63	0·38	0·57	0·47	0·05
Mn	0·19	0·16	0·25	0·26	0·52	0·15	0·18	0·16	0·15	0·19	0·41	0·41	0·42	0·31	0·12
P	0·22	0·09	0·12	0·08	0·08	0·16	0·15	0·13	0·12	0·19	0·025	—	0	0·04	0·01
K	0·09	0·07	0·07	0·09	0·09	0·03	0·14	0·04	0·06	0·13	—	0·03	0	—	0·03
Ti	0·04	0·08	0·07	0·07	0·10	0·06	0·114	0·05	0·04	0·055	0·384	0·066	0·08	0·08	0·04
Co	0·08	0·08	0·09	0·06	0·05	0·06	0·082	0·06	0·05	0·045	—	—	0·01	0·05	—
S	5·18	3·99	2·00	2·34	2·13	2·46	2·23	3·66	5·49	5·01	0·40	0·35	0·43	0·58	0·70
C	0·43	0·18	0·02	0·08	0·03	0·47	0·35	2·48	3·10	235	—	—	—	2·23	—
H	0·13	0·01	0·02	0·002	0·02	0·19	0·15	1·16	1·79	2×10^5	—	—	—	—	—

Analyses for Indarch, Ornans, Luotolax, Nova Urei and Norton County from Wiik (1969); Pillister, Cherokee Springs and Guarena from Jarosewich and Mason (1969); Leedey from Jarosewich (1967); Mokoia, Ornans, Mighei and Orgueil from Wiik (1956); Bununu from Mason (1967), and Johnstown from Mason and Jarosewich (1971). The solar values are from Ross and Aller (1976).

CO3, CO2 and C1 have the same major non-volatile element abundances and may be collectively termed CB. The apparent differences reflect differences in volatile major elements.

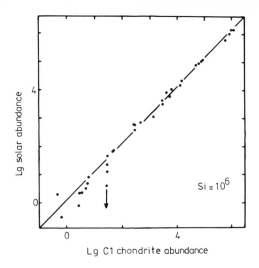

Figure 4.1. Plot comparing the abundance of 35 elements in the solar atmosphere, as determined spectroscopically, with the composition of C1 chondrites. The arrow refers to mercury, for which only an upper limit in the solar atmosphere is known. (Data from Ross and Aller 1976, Mason 1971.)

to represent material unchanged after a relatively simple formation from solar-like material.

It seems well established that the Sun formed by condensation in a gaseous cloud. The process still appears to be operative in many gaseous nebulae such as the Great Nebula in Orion, or the Trifid Nebula in Sagittarius. In these nebulae, stars which are thought to be in the final stages of formation can be identified. When the Sun formed, the same condensation process may well have given rise to the planets and meteorites. The process of condensation from a gas of solar composition explains many features of the composition of the chondrites. On the other hand, the chemistry of the achondrites, and many aspects of the chemistry of iron meteorites, are the result of processes involving melting. Before we go into detail of meteorite composition, we will discuss some of the relevant theoretical and laboratory work.

4.2. Theoretical and Laboratory Work Relating to Compositional Studies

(i) *Equilibrium Condensation.* Calculations concerning condensation in the primordial nebula have been described by

Urey (1952a), Grossman (1972) and Larimer (1973). For trace elements, the following simplified technique is often used. Condensation occurs when the vapour pressure of an element equals its partial pressure in the nebula. Vapour pressure is given as a function of temperature by

$$\ln VP(E) = -\frac{\Delta H}{RT} + \frac{\Delta S}{R}, \tag{4.1}$$

where ΔH and ΔS are the enthalpy and entropy of evaporation. By Dalton's law, the partial pressure is given by

$$P(E) = \frac{2A(E)}{A(H)} P_T(1-x), \tag{4.2}$$

where the abundances of element E and hydrogen are $A(E)$ and $A(H)$, x is the fraction condensed and P_T is the nebular pressure. This expression assumes that all the hydrogen exists as H_2, and that $P(H_2) = P_T$, since it is by far the most abundant species. Then, during condensation

$$\ln (1-x) = --\frac{\Delta H}{RT} + \frac{\Delta S}{R} + \ln \frac{[2A(E)]}{A(H)} - \ln P_T. \tag{4.3}$$

If the trace element condenses in solid solution with metal, or some other phase, the vapour pressure expression (equation (4.1)) is adjusted in accordance with Henry's law. Then

$$\ln \left(\frac{x}{1-x}\right) = \frac{\Delta H}{RT} - \frac{\Delta S}{R} + \ln \frac{[2A(Fe+Ni)]}{A(H)} - \ln \gamma + \ln P_T, \tag{4.4}$$

where γ is the activity coefficient.

The situation is slightly more complex in the case of most major elements, as they occur in a variety of compounds in both the gas and solid, but the same principles apply. The equations for every situation are set up, and solved numerically.

The results of these calculations are shown in figures 4.2 and 4.3. The first condensates are refractory minerals rich in Ca and Al, with some Mg and Si: alumina, spinel and melilite. The next important phase of condensation occurs at around 1200 K, when forsterite, enstatite and metal condense. Another major phase is sulphide, which appears at 700 K.

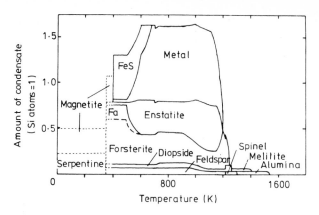

Figure 4.2. Mineral composition of the solids that condense under equilibrium conditions as a gas of solar composition cools. A nebular pressure of 10^{-6} atm has been assumed. The number of atoms, or molecules, of each mineral are expressed as a ratio relative to the number of silicon atoms in the system. Formulae for the minerals are given in table 3.3.

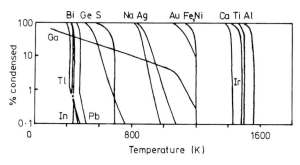

Figure 4.3. Condensation of a number of siderophile and other elements as a function of temperature.

Olivine and pyroxene are both solid solutions with Fe and Mg end members; but, until 600 K, only the Mg end members are present (forsterite and enstatite). Between 500 and 600 K, some of the metallic iron is oxidised, producing Fe end members which go into solid solution. The final stage of condensation results in the formation of magnetite, which consumes all the FeS and metal at around 400 K. It also results in the conversion of olivine and pyroxene to hydrous lattice layer silicates, like serpentine, at around 350 K. Figures 4.2 and 4.3 assume a nebular pressure of 10^{-6} atm. When higher pressures are used, condensation moves to higher temperatures—and, similarly, to lower temperatures for lower pressures—but, with a few exceptions, the sequence is unchanged. However, FeS formation and the oxidation processes are independent of pressure.

The nebula becomes increasingly oxidising as the temperature drops. It was observed above that the chondrites show progressive oxidation from top left to bottom right in figure 3.1, so that the oxidation state increases along the sequence E, H, L, LL and C. To some extent this range of oxidation can be understood in terms of the equilibrium condensation model. C1 and C2 chondrites are condensates which accreted below about 350 K. Mineralogically, they consist of magnetite and serpentine. CV and CO chondrites contain anhydrous silicates and must have accreted just above 350 K. Ordinary chondrites are somewhat less oxidised; they contain iron as metal, as FeS and as oxide in the silicates. Silicates of the appropriate Fe content must have existed in the nebula between 500 and 600 K and ordinary chondrites accreted over this range. The predicted trends in their oxidation state, and the olivine–pyroxene proportions, are similar to those observed (table 3.3), if one assumes that the order of accretion was H, L and LL. However, the E chondrites present major difficulties for an equilibrium condensation model. The required level of reduction can be found at high nebular pressures and temperatures (Herndon and Suess 1976); but, if the E chondrites accreted above 700 K, it is difficult to account for the abundance of FeS in them. We can either invoke departures from equilibrium (Blander 1971), change the composition of the gas (Larimer 1968, Baedecker and Wasson 1975), or assume a complex accretion history (D W Sears 1978 unpublished).

The condensation of many trace elements, whose behaviour can be approximated by equations (4.3) and (4.4), is shown in figure 4.3. Elements with almost upright curves with a hook at the top, condense as pure solids or compounds. The remaining curves are for elements which form solid solutions with iron or FeS. Those with an inflexion represent elements which form solid solutions, but reach their solubility limits before they are completely condensed. The remaining elements condense as pure solids. We will discuss these predictions when we come to consider trace element abundances (§ 4.3.2).

(ii) *Non-Equilibrium Condensation Models.* Infrared observations have shown that condensation and accretion occur in many nebulae at temperatures of 2000–3000 K, and in the presence of ionised gases. Arrhenius and Alfvén (1971) have proposed that such conditions in the primordial solar nebula would result in chemical fractionations similar to those observed in meteorites. For accretion, an element would need to be in the neutral state. A condensation sequence similar to that for equilibrium condensation therefore applies but is controlled by the ionisation potential instead of volatility. The main success of the model is in its explanation of why mercury is so abundant in type 6 chondrites, when similarly volatile elements (In, Tl, Bi, etc) are depleted by factors of 1000 or so. The model fails because it predicts that Ca, Al, Ba, Sr and other refractory elements, should be the last to condense. In fact, these elements are depleted in the gaseous nebulae; and this is consistent with equilibrium condensation, which predicts that they should condense first (Herbig 1974).

The model of Blander and co-workers (Blander and Katz 1967, Blander and Abdel-Gawad 1969, Blander 1971) is a modification of the equilibrium process, rather than an entirely non-equilibrium process. In this model, nucleation barriers prevent equilibrium condensation, so that considerable supersaturation of iron in the nebula occurs. The partial pressure of iron in the gas is much higher than if equilibrium were maintained. As a result, the formation of FeS and the oxidation of iron, occur at much higher temperatures (greater than 1000 K), and are pressure-dependent. In this model, the various oxidation levels observed in the chondrites are the consequence of different pressures; E chondrites formed in regions where the pressure was 10^{-1}–10^{-2} atm, and ordinary chondrites and C chondrites where the pressure was about 10^{-3} and 10^{-4} atm, respectively. Chondrules are a natural by-product of the process. In a supersaturated gas, liquids condense more readily than solids, especially at high pressures, and the chondrules would be produced by the rapid freezing of supercooled droplets.

(iii) *Melting Models.* Melting or subsequent solidification provide a number of ways of changing the composition of a system. The results are common in terrestrial rocks and, by and large, they are well understood. When a rock is completely melted the minerals may separate by density. When it is partially melted, they will separate according to their melting points: the low-melting-point minerals will melt and flow away from the others. During either process, trace elements are also redistributed because of their differing tendencies to concentrate in the liquid or solid phase. For each element in a given mineral, this tendency is expressed as the partition coefficient, k, which is the ratio of the concentration of an element in the solid, c_s, to that in the liquid, c_1. The way in which c_s and c_l change depends on the means by which the solid and liquid separated. For example, if a liquid is allowed to solidify in such a way that the solid is continuously removed from equilibrium with the liquid, the composition of the solid is given by

$$c_s = kc_o(1-g)^{k-1} \qquad (4.5)$$

where c_o is the original concentration, and g is the fraction solidified. This situation is called 'fractional crystallisation' and it may arise if the solid falls to the bottom of the liquid or sticks to the sides of the surrounding rock as it solidifies. Alternatively, the fractionation may occur by the removal of the liquid from the solid. For example, some liquid droplets may have been splashed off, or drained away, before complete solidification of the molten mass ('fractional melting' or 'partial melting'). Then

$$c_1 = \frac{1}{k}c_o(1-g)^{1/k-1}. \qquad (4.6)$$

The most common practice is to present the data on plots of one element against another, using logarithmic scales. Fractional crystallisation will then cause points to lie along a line with a slope $(k_A-1)/(k_B-1)$, for element A against element B, whilst for partial melting the slope is $(1-k_A)k_B/(1-k_B)k_A$.

(iv) *Metamorphism.* The final mechanism for fractionating the elements that we will consider is metamorphic heating. Most chondrites have undergone metamorphism involving temperatures up to about 1200 K (§ 3.2.3). The simplest theoretical treatment is to consider it as the reverse of condensation. Each element would, according to figure 4.3, be released over a very narrow temperature range as its condensation temperature was reached. This simple picture would not apply if other factors were important in metamorphism; for example, if the elements were retained at several sites, if their release were limited by low diffusion rates, or if the atmosphere during metamorphism was markedly different from that during condensation. The mechanisms by which elements are lost during metamorphism have been investigated in the laboratory by heating experiments on four chondrites (Ikramuddin *et al* 1976, 1977, Ikramuddin and Lipschutz 1976). Many volatile trace elements were lost by heating at surprisingly low temperatures. Except for the L chondrite, the amount of trace elements lost as the temperature was raised was greater at higher temperatures than below 500 °C. This showed that the loss was not from a single site by a single process. The proportions of trace elements remaining after heat treatment of the Abee E chondrite closely resembled those observed in natural E chondrites that had undergone metamorphism. These experiments therefore suggest that metamorphism may have been an important factor in the loss of trace elements from E chondrites; but there are certain problems with the results which indicate that more research is required. (For example, why were the release patterns for Tieschitz (H3) more nearly similar to the E chondrite than to Krymka (L3), another ordinary chondrite?)

4.3. Compositional Studies of Chondrites

4.3.1. Major Elements (Mg, Fe, Ca and Al)

It was pointed out above that the chondrites resemble the solar atmosphere in composition. Despite this, there are small but important differences in the major-element abundances which enable the chondrite classes to be recognised (§ 3.2.1 and table 3.2). The explanation of these chemical differences goes a long way towards explaining the formation of the classes. The problem lies mainly in converting the cosmic abundance values for the major elements, to their observed values. Let us consider first the Mg/Si ratio. This requires the removal of a Mg-rich component, or the addition of a Si-rich component. Most authors seem to favour the former. From mass–balance calculations, Larimer and Anders (1971) found that the removal of an assemblage consisting predominantly of forsterite (Mg_2SiO_4) was required. For the E chondrites, Baedecker and Wasson (1975) observed that the removal of a high-temperature condensate (melilite, spinel or alumina) with additional Mg was needed. In both instances, the Mg is removed in condensates which formed before the chondrites in question, but there is no compelling reason to believe the same mechanism applied to E and ordinary chondrites.

The variation in the Fe/Si ratio shown by the chondrites is important in distinguishing between H, L and LL chondrites, and between EI and EII chondrites. A plausible method of producing Fe/Si variations is the separation of metal and silicates. This should be very efficient, because the properties of the two phases are very different; an obvious possibility is that magnetism was involved. The abundance of Fe, and several elements which should behave similarly, are shown in the inset of figure 4.5 and under 'siderophile' in figure 4.6. The elements which condense ahead of the metal (Re, Os and Ir) show slightly greater differences between the H, L and LL classes, but those which condense later are fractionated less (Müller *et al* 1970, Ehmann *et al* 1970). This may be taken as an indication that the separation of metal and silicate occurred before the condensation of Fe. From early (incorrect) condensation calculations, Larimer and Anders (1971) found that the fractionation occurred when the temperature of the nebula passed through the Curie point of the metal. From this, it was concluded that the mechanism involved magnetism. However, subsequent calculations suggest that the fractionation occurred at 500–700 K; simultaneously with, or just before, the formation of the ordinary chondrites. The correlation between the Fe/Si ratio and oxidation state (for example, the metallic Fe/Si ratio, see table 3.2) in ordinary chondrites supports this conclusion.

The E chondrites also show a Fe/Si fractionation, but in this case the S/Si ratio varies in the same way. This may indicate that some sulphide was removed with the metal during the metal–silicate fractionation, but it then seems strange that this did not happen with the other chondrites. An alternative is that it is fortuitous: the cause of the Fe/Si variation may be independent of that responsible for the S/Si fractionation. Little work appears to have been done on the interpretation of this fractionation, although it is an essential feature of E chondrite chemistry.

The CV chondrites are distinguished from the other C chondrites primarily by their high Ca/Si and Al/Si ratios (figure 3.3). This is because they are particularly rich in aggregates of Ca–Al-rich minerals (melilite, spinel and alumina; § 3.2.2). The only time these minerals are equilibrium condensates in the solar nebula is soon after condensation begins, so the aggregates may actually be high–temperature condensates. An alternative possibility is that they are shock–melt products. The question has been satisfactorily resolved, however, by the discovery that the aggregates are enriched in refractory elements which condense at high temperatures (table 4.2; figure 4.3). If the aggregates were shock–melt products their trace elements would be governed by liquid–solid distribution coefficients, and would bear no relationship to condensation calculations. The high Ca/Si and Al/Si ratios

Table 4.2. Trace elements in the Ca–Al-rich inclusions in the Allende meteorite (abundances normalised to C1 chondrites; Wänke *et al* 1974).

Refractory elements				Non-refractory elements			
Ca	21·7	W	20·9	Co	0·093	Fe	0·047
Ti	22·2	Re	20·9	Ni	0·095	Mn	0·044
Sc	23·0	Os	17·1	Pd	0·160	Cr	0·165
V	12·7	Ir	18·4	Au	2·00	Cu	0·073
Sr	15·1	Ru	11·7			Ga	0·110
Y	20·6	Pt	11·7			Ge	0·570
Zr	22·7					As	0·890
Eu	22·8					Zn	0·23
Ta	18·7						
U	13·2						

in CV chondrites appear, therefore, to be the result of the addition of high-temperature condensates to low-temperature (CB chondrite-like) condensates (Grossman 1973, Wänke *et al* 1974).

4.3.2. Other Elements

The variations in the abundances of other elements in chondrites are presented in figures 4.4–4.6. In these figures, abundances are effectively compared with those in C1 chondrites. (This is equivalent to comparing abundances with those in the Sun, but better and more numerous data are available. The presentation chosen allows more information to be included than a simple plot of one element against another.) The simplest pattern is that shown by the C2 chondrites. The elements have been plotted in order of decreasing condensation temperatures. Most of them were chosen because they show depletion with respect to the C1 chondrites. In fact, all the elements which condense after Ni or Au (about 1400 K)

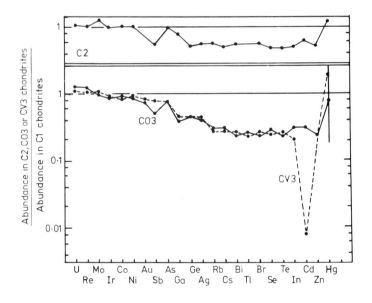

Figure 4.4. The abundance of elements in C chondrites relative to their abundance in C1 chondrites, given in order of condensation (left to right). (Data from Case *et al* 1972, Anders *et al* 1976.)

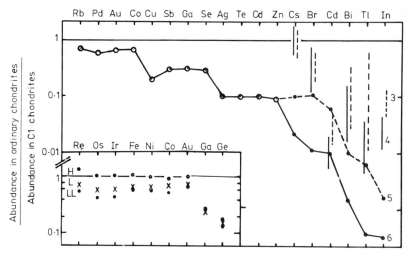

Figure 4.5. The abundance of elements in ordinary (H, L and LL) chondrites relative to their abundance in C1 chondrites, given in order of condensation. The large open circles represent elements whose abundance is independent of petrological type. The dots and vertical lines refer to elements whose abundances vary with petrological type; type 6 show lowest abundances; type 5 show a similar pattern, but the depletion is less; and types 3 and 4 show even less depletion, but greater range. The inset shows the abundance pattern of nine siderophile elements, and illustrates their different values in H, L and LL chondrites. The difference appears to be less for elements condensing later. (Data from Binz *et al* 1976, Keays *et al* 1971, Mason 1971.)

have an abundance ratio of 0·4–0·5. This appears to be related to the observation that C2 chondrites contain chondrules, while C1 chondrites do not (Larimer and Anders 1967). Chondrules may be regarded as solidified droplets formed by the melting of portions of the meteorite. Whilst molten they would have lost a number of elements by volatilisation. Volatility also governs condensation temperatures, so the elements lost from the chondrules would be those with lower condensation temperatures. Thus if the meteorites were a mixture of chondrules and matrix in equal parts, then elements missing from the chondrules should be present in only half their original proportions. This is in general accord with the observations.

The same interpretation has been applied to CV3 and CO3 chondrites, and to ordinary chondrites (Anders *et al* 1976,

100

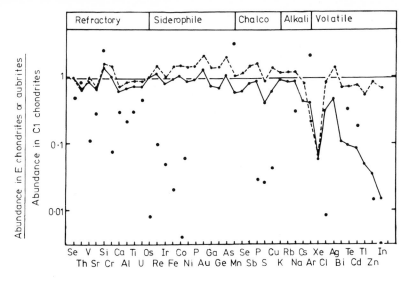

| Refractory | Siderophile | Chalco | Alkali | Volatile |

Se V Si Ca Ti Os Ir Co P Ga As Se P Cu Rb Cs Xe Ag Te Tl In
Th Sr Cr Al U Re Fe Ni Au Ge Mn Sb S K Na Ar Cl Bi Cd Zn

Figure 4.6. The abundance of elements in E chondrites relative to their abundance in C1 chondrites, given roughly in order of volatility (and therefore condensation), with geochemically similar elements together. EI (E4) analyses are connected by the broken curve and EII (E6) by the full curve. The isolated points apply to aubrite achondrites. (Data from Baedecker and Wasson 1975, Mason 1971.)

Larimer and Anders 1967), although the situation in ordinary chondrites is not clear cut. Anders and co-workers believe that there is a 0·25 plateau corresponding to meteorites consisting of 25% matrix and 75% chondrules and metal. Elements with slightly higher ratios (such as Sb and Se), Anders (1975a, 1977) argues, were only partially lost during the chondrule-forming process. Wasson and co-workers have questioned the existence of any such plateau. They suggest that the elements were fractionated by a steady loss of gas as the condensate was continually separated from the gas (Wasson and Chou 1974, Wai and Wasson 1977, Wasson 1977).

The final elements to condense—those which are the most volatile—are considerably depleted in E and ordinary chondrites, especially in the higher petrological types. Because the extent of depletion depends on petrological type, Dodd (1969b), Wood (1962) and others have proposed that these elements were all originally present in the quantities now observed in the low types, but were boiled off from the higher types by metamorphism. Larimer and Anders (1967), following a suggestion by Urey (1952a), believe that the depletion of highly volatile elements was caused by the fact that these elements were only partially condensed when accretion occurred. The higher petrological types must have accreted before the lower types, they were also buried deeper into their parent bodies and underwent greater metamorphism (assuming that the heat source was internal). This idea, if correct, enables precise values for the temperature of accretion to be estimated from figure 4.3, or element–element plots derived from it. Larimer (1973) estimates that the H, L and LL chondrites accreted over temperatures of 500–440 K, with a small systematic increase in temperature and pressure (centring on 10^{-5} atm) on going from LL to H. The proposal has been fiercely attacked by Blander (1975), mainly on the grounds of uncertainty in the thermodynamic data at these very low temperatures. We may perhaps regard them as semi-quantitative estimates.

A choice between the metamorphic and partial-condensation models for explaining the highly volatile element patterns should be possible from laboratory heating experiments and cooling rate estimates. Heating experiments on little-metamorphosed ordinary chondrites argue against the metamorphic model (§ 4.2). Cooling rates are governed by burial depths, so that their relative values are an indication of accretion order. Unfortunately, they are ambiguous, since estimates based on charged-particle tracks favour the partial accretion model, while the metallographic estimates favour metamorphism.

The major features of the chemistry of the E chondrites centre on the differences between the EI and EII chondrites; in particular, the variations in the abundance of Fe (which is the same for the other siderophiles), and S (and the other chalcophiles). These were discussed earlier. The interpretation of the depletion in highly volatile elements in the higher petrological types, is subject to the same dispute as for the ordinary chondrites. In this case, however, the heating experiments support the metamorphic model.

102

4.4. Compositional Studies of Achondrites

With the possible exception of the aubrites, the achondrite classes are igneous in origin and their compositions can be interpreted by melting models. The ureilites seem to be the solid residue left behind when some chondrite-like material was partially melted (Ringwood 1960a). Table 4.1 shows that these achondrites are low in Fe, Ni, S, Na and Al; elements which, in chondrites, occur in low-melting-point minerals (metal, sulphide and feldspar). The ureilites could therefore be removed if the parent material were heated to a temperature sufficient to melt these minerals, but not the olivine and pyroxene. The chemistry of the ureilites is complicated by the ubiquitous carbonaceous veins, which form about half of the meteorite. Trace element studies face the major problem of disentangling different trends in the two materials: it seems that the veins are richer in metal than the matrix, but are very poor in volatile elements except rare gases and carbonaceous compounds. The veins were probably injected into the ureilite parent rock at a late stage in its history (Kenna 1976, Weber *et al* 1971).

The eucrites and howardites are enriched in Ca, Al and other elements; that is, predominantly low-density minerals. These would float to the top if a body resembling the chondrites were melted and, after cooling and solidifying, a crust resembling the eucrites and howardites would form. Since they are also low-melting-point minerals, a partial-melting mechanism cannot be ruled out. McCarthy *et al* (1973) have interpreted the Sr and Zr distributions in terms of the former mechanism. Their results are summarised in figure 4.7. If pyroxene, with Sr and Zr abundances equivalent to position A were melted, and fractional crystallisation allowed to take place, equation (4.5) predicts that the solids would lie along the 'pyroxene fractionation' line. Similarly, fractional crystallisation of a plagioclase melt would give the 'plagioclase fractionation' line. The melt is an approximately equal mixture of both minerals, so the resulting trend would be similar to that observed. In fact, the parent melt need not be at point A in the figure: it is sufficient that the material can produce Sr and Zr abundances equivalent to point A. Certain of the howar-

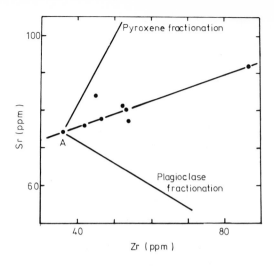

Figure 4.7. Strontium and zirconium abundances in seven eucrites. The two fractionation lines indicate the change caused by complete melting of a meteorite with composition A followed by fractional crystallisation. The partition coefficients for Sr in pyroxene and plagioclase were taken to be 0·08 and 0·001, and the coefficients for Zr in pyroxene and plagioclase were taken to be 1·5 and 0·001, respectively. Since these meteorites are an approximately equal mixture of the two minerals, the observed trend is predicted (McCarthy *et al* 1973).

dites are likely candidates; but, to lower the amount of Mg (see table 4.1), some olivine or pyroxene would need to be removed. These high-density minerals would fall to the bottom of a magma chamber, and the resulting 'cumulates' would closely resemble another achondrite class, the diogenites. The rare earth element studies are consistent with this model. The uniform enrichment of these elements in eucrites and howardites is similar to that in terrestrial rocks which are known to have been completely melted (Schnetzler and Philpotts 1969). However, Gast and Hubbard (1970) have also shown that the abundances can be explained by partial melting. Stolper (1977) has recently pointed out that the FeO content of the eucrites is slightly higher than could be produced by a chondritic melt. It seems that some addition of iron is also required. The howardites are highly brecciated, sometimes with eucrite and diogenite fragments mixed together. McCarthy *et al* (1972) have consequently proposed that they are mechanical mixtures of eucrites and diogenites.

It is generally assumed that the ultimate parent material for these achondrites was chondritic but, by and large, the melting process has completely obliterated the effects of condensation and accretion. The highly volatile elements, which show 1000-fold variations in abundance in ordinary chondrites, have similar abundance patterns to oceanic basalts. Laul *et al* (1972) have interpreted this to mean that they had similar accretion temperatures.

The aubrites appear to be the only achondrites which may not be the products of a melting process. Compositionally and mineralogically, they resemble the E chondrites, except in their much lower iron and sulphur contents. In fact, Wasson and Wai (1970) found that they formed a continuous sequence with the E chondrites in their metal, schreibersite and perryite compositions, and in their Mn/Si and Na/Si ratios. The aubrites have been exposed to cosmic radiation for longer than EII chondrites, which, in turn, have been exposed for longer than EI chondrites (figure 5.8). Aubrites are also rich in gases implanted by the solar wind, which implies a surface location (§ 4.7.1). Wasson and Wai (1970) therefore conclude that these meteorites originated from a common parent body which contained EI at its centre, EII at intermediate depths, and aubrites on its surface. The scheme places the most intensely metamorphosed and recrystallised meteorites at the surface, so that an external heat source is required; early intense solar radiation and intense meteoritic bombardment are possibilities for this. Prior to Wasson and Wai's data becoming available, it was usually assumed that the aubrites were igneous. This was primarily because of their coarse texture. The complex, and it now seems unlikely, igneous processes which could produce the aubrites have been reviewed by Mason (1962).

4.5. Compositional Studies of Stony-Irons

The most comprehensive study of pallasite composition is probably that by Scott (1977): 19 out of 28 pallasites form a group in which Ni correlates negatively with Ga and Ge, and positively with Au. All the trace elements studied are present in similar abundances to the Ni-rich end of the IIIAB iron meteorites. Scott (1977) therefore proposed that the pallasites are material from the core–mantle boundary of their mutual parent body. Urey (1966) had previously suggested that they came from the edges of iron bodies imbedded in a stony body. Trace element analysis has also led to the separation of the Eagle Station, Itzawisis and Cold Bay meteorites from each other (cf § 3.4.2; table 5.3).

Wasson *et al* (1974) have examined the Ni, Ga, Ge and Ir contents of metal from the mesosiderites (table 4.3). These

Table 4.3. Composition of metal in stony-iron meteorites (Wasson *et al* 1974, Scott 1977).

	Number analysed	Ni (%)	Ga (ppm)	Ge (ppm)	Ir (ppm)	Au (ppm)
Main pallasite group	19	7·8–11·7	16–26	29–65		1·7–3·0
Other pallasites†	3	14–16	4·5–6·0	75–120		0·8–1·0
Mesosiderites	17	7·0–9·0	13–16	47–58	2·4–4·4	

† Eagle Station, Springwater and Itzawisis, which stand apart from the main pallastie group but resemble each other.

elements are generally present in the same abundances as they are in IIIAB, IIE irons and the pallasites. Correlations exist between Ga–Ge, Ni–Ge and Ni–Ga, and the IIE meteorites lie on an extrapolation of these trends. This therefore suggests a link between the IIE irons and the mesosiderites.

4.6. Compositional Studies of Iron Meteorites

4.6.1. *Compositional Differences Between the Groups and their Formation*

Iron meteorites are essentially iron–nickel alloys, and contain small amounts of cobalt, phosphorus, carbon and sulphur.

106

Table 4.4. Major- and minor-element abundances (wt %) in the major iron meteorite groups (iron constitutes the balance)†

Group	Ni	Co	P	C	S Chemical	S Planimetric
IAB	7·53	0·46	0·18	0·04‡	0·004	1·64
IIAB	5·73	0·43	0·26	0·006	0·006	0·17
IIIAB	8·27	0·44	0·15	0·010	0·004	0·77
IVA	8·23	0·39	0·05	0·010	0·006	0·21
IVB	16·8	0·65	0·05	0·004	0·006	0·03

† Data from Moore *et al* (1969) and Lewis and Moore (1971), except Ni values which are from Wasson (1974), and the planimetric S values from Buchwald (1974).
‡ Mean is poorly defined, as values are evenly spread over the range 0·005–0·06, and go up to 0·46.

Abundances in the major groups (§ 3.5.1) are given in table 4.4. The determination of the amounts of sulphur, carbon and, to a lesser extent, phosphorus is particularly difficult, because these elements occur in large, sparsely distributed inclusions which it is impossible to sample fairly. Mineralogical observations are therefore essential (table 3.9). In some instances, estimates can be made from the surface area of a section occupied by their minerals (the 'planimetric method'). The two sets of values for sulphur give an indication of the size of the problem; they differ by one to two orders of magnitude. Despite these difficulties, we can see certain trends; phosphorus is the dominant minor element and its concentration is much the same in all group averages, being only slightly higher in IIAB and slightly lower in IVA and IVB, than in IAB and IIIAB. Carbon is far more abundant in IAB than in the others, and IVB contains much less sulphur than the other groups.

The iron meteorites are classified on the basis of their Ni, Ga and Ge content. A condensation–accretion model of some kind is probably capable of explaining these groups. Wasson and Wai (1976) consider the low Ga and Ge abundances in IVA and IVB to be a consequence of an aerosol, rich in these elements, being blown away by the solar wind. Kelly and Larimer (1977) assume that all the iron meteorites were formed at the same nebular pressure, with their Ga and Ge content resulting from accretion at different temperatures. The variation in Ni content, Kelly and Larimer argue, is due to a discrete, subsequent oxidation event within the parent body. Sears (1978) considers that the oxidation process is part of the overall condensation process, as it appears to have been for the chondrites. In this case, except for IVB, the groups formed at much the same temperature (600–700 K), but over a wide range of pressures: 10^{-4} atm for IAB and IIAB (with a small temperature difference), 5×10^{-6} atm for IIIAB, and 10^{-8} atm for IVA. This model also explains the distribution of carbon and sulphur. Only groups which accreted below 700 K (the formation temperature of FeS) would be expected to contain sulphur, whilst the presence of significant amounts of carbon in only one group is consistent with the very special pressure–temperature conditions needed for its condensation.

The small IVB group appears to have been produced under quite different conditions (Scott 1972, Kelly and Larimer 1977). Its Ga, Ge and Ni contents can be found in the first metal to condense, provided a pressure of less than 10^{-3} atm is assumed. Such metal is enriched in nickel, because nickel condenses faster than iron (its vapour pressure is lower). The lower sulphur abundance in IVB results from its accretion before this element had condensed (700 K). Figure 4.8 shows the abundance of a number of other trace elements in a IAB and IVB iron meteorite, compared with C1 chondrites. The IAB iron has most elements present in the same proportions as C1 chondrites. Other classes are probably similar to IAB in this respect, but comparison is difficult because the ranges within the groups are very large. In contrast, the IVB meteorite is enriched in elements condensing before Fe and Ni, and depleted in those condensing afterwards. This seems to be good confirmation of the accretion conditions proposed above. IVB accreted before the elements more volatile than iron had condensed, and they are therefore depleted. The other classes condensed at such a low temperature that most elements, except the particularly volatile Ga and Ge, were fully condensed. Hence they are present in their original (C1) proportions.

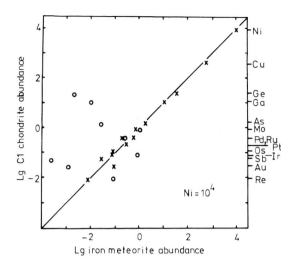

Figure 4.8. Plot comparing the abundance of siderophile elements in C1 chondrites with two iron meteorites. Crosses represent the Canyon Diablo IAB meteorite which has these elements in the same proportion as C1 chondrites, and open circles represent the Hoba IVB meteorite which is depleted in volatile elements (Ga, Ge, As, Sb and Au) and enriched in refractory elements (Ir and Re). (See also figure 4.9. Data from Smales *et al* 1967, Mason 1971.)

4.6.2. Compositional Trends Within the Groups

Figure 4.9 presents two element–element plots for iron meteorites. In groups IIAB, IIIAB and IVA (there are too little data for group IVB) Ir shows negative, and Au positive, correlations with Ni. Most trace elements show similar behaviour to one of these elements (Scott 1972).† The correlations seem to be the result of a melting process, but it is not clear whether they are due to partial melting (Kelly and Larimer 1977) or fractional crystallisation (Lovering 1957, Scott 1972). For example, the line through the Ir–Ni data for group IIIAB in figure 4.9 could represent fractional crystallisation (where k_A and k_B are 3·0 and 0·92, respectively), or partial melting (where these values are 3·0 and 0·97). On our present knowledge of partition coefficients, a choice cannot be

---

† Notable exceptions are the Ge–Ni and Ga–Ni plots in IIIAB, which change slope so that correlations in the low-Ni IIIA end are positive, and the high-Ni IIIB end are negative.

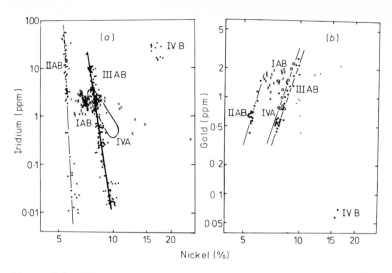

Figure 4.9. Element against nickel plots for (*a*) iridium (data from Wasson 1974) and (*b*) gold (Scott 1972). Most of the refractory elements show negative correlations with nickel, like iridium, and most of the volatile elements show positive correlations, like gold. The significance of the trends and, in particular, the bold curve through IIIAB for Ir–Ni, are discussed in the text.

made between these two melting models. Group IAB stands out in that it does not show strong element correlations; this is especially true for the refractory elements such as Ir. Scott (1972) suggested that this group had escaped the melting episode; whereas Kelly and Larimer (1977) argued that IAB irons are liquids from a fractional melting process. The presence of silicates in certain IAB members—and also in IIE irons and many anomalous meteorites—has suggested to several authors that these irons have not been completely melted (Wasson 1970, Wasserburg *et al* 1968, Urey 1966).

4.7. The Inert Gas Content of Meteorites

4.7.1. Solar and Planetary Trapped Gases

We saw in § 1.5.4 that the inert gases in meteorites appear to be essentially of three kinds: (i) those produced by cosmic ray bombardment; (ii) those resulting from radioactive decay of elements in the meteorite, and (iii) an amount which was present originally (the 'trapped' or 'primordial' gases). The

110

last component will be considered here. The trapped gases—
^{4}He, ^{20}Ne, ^{36}Ar, ^{84}Kr and ^{132}Xe—come from a variety of
sources. One source contributes each inert gas in essentially
solar proportions (and may be conveniently referred to as
'solar-type' gas), whilst in the other, the relative proportions
are those of the Earth's atmosphere ('planetary-type' gas)
(Signer and Suess 1963). The relative amounts of each type of
gas may be determined by considering the isotopes of a single
element; for example, the three isotopes of Ne with atomic
numbers 20, 21 and 22. ^{21}Ne results almost entirely from
cosmic ray bombardment, but we will use ^{22}Ne as the standard
on which to base our measurements (ratios are far easier to
determine accurately than absolute amounts). The ^{20}Ne/^{22}Ne
ratio for a large number of meteorites is plotted against
^{21}Ne/^{22}Ne in figure 4.10. Because there are three
components, the points plot over a triangle. The plot with

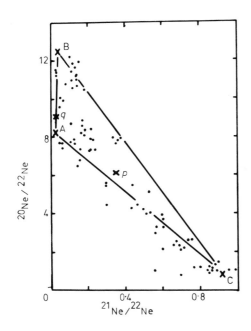

Figure 4.10. Three-isotope plot for Ne in gas-rich meteorites.
The point C corresponds to cosmogenic neon, and points A and
B correspond to two types of trapped neon. Such a plot enables
the contribution from each source to be estimated. For example, for a gas plotting at point p, the proportion of cosmogenic
gas to total gas in Cp/Cq, and the proportion of Ne–A in the
trapped gases is Aq/qB (Heymann 1971).

high ^{21}Ne is mainly cosmogenic, and the other two are the trapped components. The proportion of each component in the gases from a given meteorite is proportional to the distance it plots from the corresponding apex of the triangle in figure 4.10. Some meteorites contain only solar-type gas, and some only planetary-type trapped gases, whilst many contain a mixture.

Black and Pepin (1969) have pointed out that ^{20}Ne/^{22}Ne values sometimes occur well outside the triangle in figure 4.10, when the gases released at various temperatures are examined separately. This suggests the need for additional components corresponding to a ^{20}Ne/^{22}Ne ratio of less than 3·4 (Ne–E; C chondrites only) and greater than about 14 (Ne–D). In addition, the gases released over certain temperature ranges lie on a line between the cosmogenic point and a ^{20}Ne/^{22}Ne ratio of 11.2, which suggests another source corresponding to this value (Ne–C). Black (1972a,b) has been able to determine He and Ar isotope ratios for these different components (table 4.5). He suggests that component C is solar-flare-implanted gases, component D is rare gas ions implanted by the pre-main-sequence solar wind, and component E is extra-solar-system gases.

Table 4.5. Isotopic composition of the trapped gases in C chondrites and other gas-rich meteorites (Black 1972a,b, Heymann 1971).

	^{3}He/^{4}He $(\times 10^4)$	^{20}Ne/^{22}Ne	^{21}Ne/^{22}Ne	^{36}Ar/^{38}Ar
Component A†	1·7	8·2	0·025	5·21
Component B‡	3·9	12·52	0·0335	5·37
Component C	4·1	10·6	0·042	4·1
Component D	1·5	14·5	—	6·0
Component E	—	3·4	—	—

† Planetary-type.
‡ Solar-type.

4.7.2. Gas-Rich Meteorites

According to present studies, all of the C chondrites and 12, 1·7, 8, 33 and 33% of the H, L, LL, howardite and aubrite groups, respectively, are gas-rich (Schultz *et al* 1971, 1972). Except for the C chondrites, the gas-rich meteorites are breccias consisting of xenoliths ('discrete fragments') of the normal, comparatively light colour, in a matrix of dark material: this is usually referred to as the light–dark structure. Gas data for one of the L chondrites are presented in table 4.6.

Table 4.6. He, Ne and Ar abundances in the Assam L5 chondrite (10^{-8} cm^3 g^{-1} STP) (Schultz *et al* 1971).

	^{3}He	^{4}He	^{20}Ne	^{21}Ne	^{22}Ne	^{36}Ar	^{38}Ar	^{40}Ar
Light portion	58·1	1663	9·03	8·65	10·25	2·02	1·77	5160
Dark portion	86·9	78300	169·0	9·43	23·80	6·98	2·67	6500

The dark portion is enriched in every isotope, but especially in ^{4}He and ^{20}Ne. ^{4}He is normally considered to be a radiogenic isotope, but only about 2000×10^{-8} cm^3 g^{-1} STP could be accounted for in this way. When the Ne isotopes are plotted (figure 4.10), those in the light portion are found to be almost entirely cosmogenic. The dark portion, however, contains considerable quantities of solar-type trapped Ne. It is the abundance of solar-type gas which causes these meteorites to be gas-rich.

4.7.3. Non-Gas Rich Ordinary Chondrites

Ar, Kr and Xe are present in ordinary chondrites as trapped gases, and trapped Ne has been observed in some petrological type 3 chondrites (Kirsten *et al* 1963, Zähringer 1968, Heymann 1971). The abundance of these elements correlates with petrological type (figure 4.11). In this property Ne resembles other highly volatile elements, and the same arguments exist concerning its cause (§ 4.3.2). It seems probable that the trapped gas in non-gas-rich ordinary chondrites is solar-type (Heymann 1971).

4.7.4. The Origin of Trapped Gases in Meteorites

The origin of the solar-type gases now seems well established. Etching experiments (Eberhardt *et al* 1965a, b, Hintenberger

113

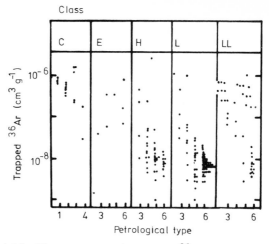

Figure 4.11. The amount of trapped ^{36}Ar in each chondrite class for each petrological type (Zähringer 1968).

et al 1965), and changes in the release patterns under electron beam bombardment in the microprobe (Zähringer 1966), have shown that the gases are located on the surface of grains. This supports the early suggestions that the gases are implanted by the solar wind (Suess *et al* 1964, Signer 1964, Wänke 1965). They are present in the dark portions of meteorites with light–dark structure. At some time the dark material has been exposed to the solar wind, presumably as the surface regolith on the meteorite parent body.

The origin of the planetary-type gases is still unclear. Diffusive loss, proposed by Zähringer (1962), was not responsible since the fractionation is no greater for the lighter isotopes. Other mass-dependent mechanisms run into similar difficulties (Kuroda and Manuel 1970). Mechanisms depending on different physical properties, such as adsorption onto silicates (Mazor *et al* 1970) and solubility in magnetite (Lancet and Anders 1973), seem more promising, but silicate adsorption requires implausibly low temperatures (Fanale and Cannon 1972).

4.7.5. Xenon and Krypton

The abundance pattern of Xe isotopes in C chondrites is unique, and the acronym AVCC (average carbonaceous chondrite) Xe has been adopted for it. The light isotopes (^{124}Xe,

^{126}Xe and ^{128}Xe) in the C chondrites are present in the same proportions, when normalised to ^{130}Xe, as they are in the Sun. The heavy isotopes (^{132}Xe, ^{134}Xe and ^{136}Xe) are more abundant in the C chondrites. Reynolds and Turner (1964) have suggested that the heavy isotopes are fission products, but Eugster *et al* (1967) could find no fissile nuclide capable of giving the observed pattern. Anders and Heymann (1969) have suggested alternatively that the parent nuclide was a now-extinct superheavy element, whilst Kuroda (1971) has proposed that nuclear reactions in the early Sun were responsible for the excess heavy Xe. The abundance pattern for the Kr isotopes is also difficult to interpret, and seems to require a fissionogenic component (Eugster *et al* 1967, Marti 1967).

4.8. The Organic Constituents of Meteorites

4.8.1. Carbon Content and Distribution

The abundance of carbon in certain meteorites is considered to be of special significance for a variety of reasons. In the first instance, its presence testifies to the primitive nature of C1 and C2 meteorites (§ 3.2.2). Secondly, it seems relevant to the origins of life on Earth. There is no obvious way to condense carbon in the primordial solar nebula by equilibrium processes operating at any reasonable pressure and temperature. Explaining its presence in meteorites amounts to explaining its presence on Earth. Thirdly, many interstellar molecules are carbon compounds. They seem to have formed in an environment similar to that of the meteorite hydrocarbons, and a similar mode of synthesis is probable.

The meteorites richest in carbon are the C chondrites—which may contain up to 5%—and the ureilites (table 4.1). The form in which the carbon is present in seven C chondrites is indicated in table 4.7. About 5% is present in the form of organic compounds which can be removed by refluxing with a solvent (in this case with a benzene–methanol mixture), and a similar amount is present as carbonate. About half the carbon (in the case of Mokoia 63%) is a compound which is insoluble in organic solvent and acid. (Its only means of isolation is by

Table 4.7. Distribution of carbon in seven C chondrites (Smith and Kaplan 1970).

| Chondrite | Class | Carbon content (wt%) | | | |
		Carbonate	Insoluble portion	Soluble portion	Total†
Ivuna	C1	0·20	4·11	0·07	4·03
Orgueil	C1	0·13	2·15	0·11	3·75
Mighei	C2	0·21	0·72	0·09	2·85
Cold Bokkeveld	C2	0·07	2·98	0·11	2·35
Erakot	C2	0·05	2·43	0·13	2·30
Murray	C2	0·13	2·78	0·11	2·24
Mokoia	C3	0·00	1·08	0·06	0·74

† The difference between this value and the sum of the previous three values reflects experimental error.

dissolving the rest in hydrofluoric acid.) In ureilites, half the carbon is present as diamonds which, as in the Canyon Diablo IAB iron, are of shock origin. In other classes, it exists as carbides (IAB irons and E chondrites) and graphite (all except C1 chondrites).

4.8.2. Contamination and Isotope Ratios

The problem of contamination is illustrated most graphically by the finding of Oro and Skewes (1965) that the quantities of some constituents present (the amino acids) are equal to the amounts that would be transferred to the meteorite in ten fingerprints! Specimens have been in museums for up to 170 years, so the danger of contamination from handling, and from airborne organics, is considerable. To make things worse, it is likely that some fragments will have lain on the surface of the Earth for several weeks before collection, and this is long enough for serious contamination. (One of the Lost City fragments was contaminated with dog urine within minutes of falling.)

The discovery of 'organised elements' caused considerable excitement a few years ago. These small bodies, generally less than about 1 mm in diameter, had complex structures, but were not crystalline. They appeared to be biological, and this led to the resurrection of the idea that early life forms had been carried to the Earth on meteorites. However, these bodies are also largely due to contamination. The more complex structures have been identified as a variety of pollen which had obviously impregnated the meteorite during its terrestrial sojourn. It seems probable that the simpler organised elements are also contaminants, but this is unproven. A number of organised elements have been identified as mineral grains (Vdovykin 1967). Nagy (1975) has written a lengthy review on this topic.

Fortunately, there is evidence that at least a major fraction of the organic compounds are indigenous. This is provided by the carbon, hydrogen and sulphur isotope ratios (Briggs and Mamikunian 1963, Smith and Kaplan 1970). The isotope ratios of the carbon in the constituents of seven C chondrites is given in table 4.8. In terrestrial organic compounds $\delta^{13}C$ values lie between -15 and -60 (usually -25 to -35). The insoluble organics and the carbonates are clearly indigenous. The soluble organics may contain terrestrial contamination, but Orgueil, Cold Bokkeveld and Murray have such similar

Table 4.8. Isotopic abundances of ^{13}C (as $\delta^{13}C$)† for meteorite fractions relative to the Pee Dee belemite standard (Smith and Kaplan 1970).

Chondrite	Carbonate	Soluble organics	Insoluble organics
Ivuna	+65·8	−24·1	−17·1
Orgueil	+70·2	−18·0	−16·9
Mighei	+42·3	−5·3	−14·8
Cold Bokkeveld	+50·7	−17·8	−16·4
Erakot	+44·4	−19·1	−15·1
Murray	+41·6	−17·8	−16·8
Mokoia	none	−27·2	−15·8

† The ratio of the two isotopes of carbon is conveniently expressed in terms of $\delta^{13}C$, which is defined by

$$\delta^{13}C = \frac{(^{13}C/^{12}C)_{sample} - (^{13}C/^{12}C)_{standard}}{(^{13}C/^{12}C)_{standard}}.$$

The standard is a universally agreed standard rock. The units of $\delta^{13}C$ are per mil (‰), and are positive if the sample is enriched in ^{13}C with respect to the standard.

values, that it seems more probable that these three are uncontaminated, and that the others contain varying proportions of contamination.

Why the carbon isotope ratios should be distinct from those observed on Earth, especially in the carbonates, is a problem. Anders *et al* (1973) have argued that the distinction is due to the kinetic isotope effect during synthesis of the organics, whilst Arrhenius and Alfvén (1971) have argued that it is due to the condensation mechanism.

4.8.3. *The Insoluble Organic Portion*

X-ray diffraction studies have shown that the insoluble organic material is neither graphite nor amorphous carbon. It is therefore either a polymer, or organic compounds trapped in the crystal structure of the silicates. If the latter were the case, pyrolitic release experiments would yield the same patterns as for the soluble organics. Since they do not, carbon is generally assumed to be present as a polymer (Bandurski and Nagy 1976, Hayatsu *et al* 1977).

The polymer appears in the form of 2–3 μm translucent flakes mixed with equi-sized silicate grains (Vdovykin 1967). Infrared spectroscopy and electron spin resonance have indicated the presence of CO and OH groups and an aromatic skeleton (Hayes 1967). As an example of the kind of data available from pyrolitic release experiments the work of Bandurski and Nagy (1976) may be mentioned. 42 pyrolysis products were recognised with three types of release pattern: those which are released at low temperatures (e.g. thiophene); those which are thermally stable and produced only at high temperatures (e.g. benzothiophene); and some which are intermediate (e.g. xylene). The most readily released fragments are branches, or bridges, in the polymer. The least labile probably represent the main skeleton. An idea of the complexity of the structure is apparent in the authors' conclusion that their earlier idea—that the polymer consisted of a 'condensed aromatic and heteroaromatic matrix with aliphatic bridges and sidechains'—may be an oversimplification. Briggs and Mamikunian (1963) likened the polymer to humic acid in soils.

4.8.4. The Soluble Organic Portion

The most abundant family of soluble organic compounds seems to be the alkanes (C_nH_{2n+2}) (Meinschein 1963, Studier *et al* 1972). The most common alkane probably has n near 15 (Hayes 1967), but it is difficult to be certain. Most alkanes are straight chains, and only a few are branched. Loss during solvent extraction depletes the low alkanes, and contaminants are usually heavy alkanes. Attempts have been made to avoid loss during extraction by using sealed vials, and this has proved very successful in locating highly volatile organics (Jungclaus *et al* 1976). Studier *et al* (1968) found that for compounds with n below 11, aromatic hydrocarbons (that is, hydrocarbons based on benzene rings) are more common than alkanes.

Amino acids ($NH_2C_nH_{2n}COOH$) are important because they are essential to the synthesis of proteins and the precursor molecules to life on Earth. Wolman *et al* (1972) have listed 18 amino acids found in meteorites, the most abundant being glycine ($n = 1$) and alanine ($n = 2$). Their list includes both protein and non-protein members.

Monocarboxylic acids (RCO_2H, where R stands for any organic radical) and dicarboxylic acids ($R(CO_2H)_2$) have been found by most authors (for example, Lawless *et al* 1974). Smith and Kaplan (1970) thought they represented contamination. Similarly, pristane and phytane (which have been found by several authors, including Smith and Kaplan), have been dismissed as contamination by Anders *et al* (1973). The number of compounds which have been found and disputed is considerable. They include a number of chlorine- and sulphur-bearing constituents.

4.8.5. The Origin of the Organic Compounds

Terrestrial organics derived from biological sources are always optically active. This is not true of meteoritic organics (Hayatsu 1966), despite claims to the contrary (Nagy *et al* 1961), and the organic constituents of meteorites appear to be the result of non-biological processes. There are two rival theories to account for the formation, both of which were originally postulated by Urey. The first theory is that they resulted from the effect of energetic radiation (either from the

119

Sun, or from lightning flashes in the primordial nebula) on a gaseous mixture of H_2O, CH_4 and NH_3 (Urey 1952b). This situation was successfully simulated in the laboratory by Miller (1953), and is generally termed the Miller–Urey synthesis. The second is that, in the presence of a catalyst, CO may react with hydrogen to produce hydrocarbons (Urey 1953). The presence of a small amount of NH_3 would result in the nitrogen-bearing compounds (Hayatsu *et al* 1968). This is the well known, and commercially important, Fischer–Tropsch reaction.

Synthesis experiments using both methods have been carried out in the laboratory (Anders *et al* 1973, Wolman *et al* 1972). Both methods give alkanes, carboxylic acids, amino acids and polymers. However, detailed consideration suggests that the Miller–Urey synthesis is to be preferred for amino acids, but that the Fischer–Tropsch synthesis is better for alkanes. A major difficulty of the Fischer–Tropsch-type mechanism is that it requires carbon to be present in the nebula in the form of CO, whereas equilibrium calculations suggest that it should be present as CH_4 at the temperatures corresponding to meteorite formation. Furthermore, the Fischer–Tropsch mechanism can produce amino acids only after prolonged periods at high temperatures, and these would have destroyed many of the low-temperature phases in the meteorites. The question is still unsettled, and it may be that a combination of processes is required.

5. Physical Properties and Processes

5.1. Physical Properties

5.1.1. Mechanical Properties

The densities, porosities, compressive strengths and seismic velocities have been compiled for a number of meteorites by Wood (1963). Some of his data have been updated by Wasson (1974). The density of a meteorite depends mainly on its metal content. Iron meteorites have densities between 7·70 and 7·90 g cm^{-3}. Those which have been cosmically reheated have somewhat lower values (7·6–7·7; Henderson and Perry 1954). Most stony meteorite densities are 3·5–3·8 g cm^{-3}; whilst for metal-free C chondrites the values are 2·2–2·9 g cm^{-3}. Porosities and seismic velocities are generally similar to those of terrestrial sandstones (1–20% and 3000–4000 km s^{-1}, respectively). They show no correlation with petrological type or meteorite class. Compressive strengths are typically 1000–4000 kg cm^{-2} for ordinary chondrites, but no doubt they would be somewhat lower for C chondrites.

The fracture mechanism of iron meteorites has recently been much discussed. Gordon (1970) has suggested that the scarcity of pre-terrestrial deformation in iron meteorites is inconsistent with the theory of their origin by fragmentation of metallic cores. This conclusion was based on the assumption that during its fragmentation the body was heated sufficiently to make the metal ductile. It has been challenged by Johnson and Remo (1974) who have argued, instead, that it inferred that the body was small (less than 140 km in diameter). In this case the metal would stay cool enough to remain brittle. However, a wide variety of fracture mechanisms are possible (parting along grain boundaries and crystal faces, for example); so neither of these conclusions can be accepted without reservation (Marcus and Palmberg 1971, Marcus and Hackett 1974).

5.1.2. Thermal Properties

Tabulations of thermal conductivity and specific heat have also been made by Wood (1963). Both are related to the metal content of the meteorite; and thermal conductivity is also affected by porosity. Values are typically $0 \cdot 2$ cal g^{-1} °C and 4×10^{-3} cal s^{-1} cm °C for specific heat and thermal conductivity, respectively.

5.1.3. Optical Properties (Reflectivity and Luminescence)

Reflectivity measurements usually extend from $0 \cdot 3$ to around $2 \mu m$, to they cover the near-ultraviolet, visible and near-infrared wavelengths. There are usually two main absorption bands. One is very intense and appears at $0 \cdot 3 \mu m$. The other varies in intensity, and occurs at $0 \cdot 9 - 1 \cdot 0 \mu m$. This is most useful in meteorite studies, and is due to Fe^{2+} in pyroxene. The exact position and intensity is determined by the proportion of Fe end member in the pyroxene. (Fe^{2+} in olivine causes weaker absorption at $1 \cdot 1 \mu m$.)

Figure 5.1 shows examples of the reflectivity spectra of four meteorites. Allende (CV3) and Indarch (E4), in contrast with Bachmut (L6) and Kapoeta (howardite), do not contain Fe-bearing pyroxene. As a consequence, their spectra have no $0 \cdot 9 - 1 \cdot 0 \mu m$ absorption. Table 5·1 illustrates how the depth and location of this absorption band relate to the amount of Fe-pyroxene (cf table 3·3).

The absolute reflectivity (albedo) at $0 \cdot 56 \mu m$ of the ordinary chondrites correlates to some extent with petrological type: $12 \cdot 5 - 16 \cdot 5\%$ and $18 \cdot 5 - 24 \cdot 0\%$ for types 3 and 4, respectively, and $20 \cdot 5 - 33 \cdot 0\%$ for types 5 and 6 (Chapman and Salisbury 1973). These values are higher than for C chondrites, which are 3–5% for C1 and C2 chondrites, and 5–15% for C3 chondrites (Johnson and Fanale 1973). Some polarisation measurements have been made on Bruderheim (L6) by Egan et al (1973).

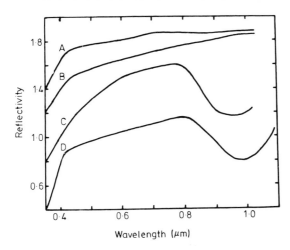

Figure 5.1. Reflectivity spectra of four meteorites. The reflectivity scale applies to Kapoeta (a Ca-rich achondrite), curve D, which has been set equal to unity at 0·56 μm. The other curves have been displaced for clarity: A, Allende (CV3); B, Indarch (E4), and C, Bachmut (L6) (Chapman and Salisbury 1973).

Table 5.1. Spectral reflectivity of meteorites (Chapman and Salisbury 1973).

Group	Number of specimens	Location of absorption band (μm)	Reflectivity at 1·1 μm / Reflectivity at 0·75 μm
H	4	0·93 (0·925–0·935)	1·00 (0·97–1·04)
L	9	0·95 (0·935–0·960)	0·91 (0·79–0·98)
LL	4	0·97 (0·955–0·980)	0·87 (0·83–0·90)

Reflectivity measurements have acquired particular interest lately, as they provide a means of comparison with the asteroids, which may possibly be the parent bodies of meteorites. Most of the main-belt asteroids have reflectivity spectra resembling C chondrites, although they could equally well be opaque-rich basalts (Chapman 1976). Only two asteroids out of the 100 so far measured have spectra resembling those of the ordinary chondrites, and one of these is an Apollo asteroid (Chapman 1976, Chapman and Salisbury 1973; § 6.3.5). (Apollo asteroids may be unrelated to the main asteroid belt.) A notable match is between the Nuevo Laredo eucrite and Vesta (McCord *et al* 1970), which is one of the largest asteroids. Unlike most of the asteroids, which are the result of multiple fragmentation, Vesta is probably still in its original form (Dohnanyi 1971). This work therefore suggests that C chondrites and eucrites may come from the main asteroid belt, but the ordinary chondrites may not. However, in the absence of density data these findings must be treated with caution; they could apply only to a thin surface layer.

Studies of meteorite luminescence were originally prompted by the suggestion that the transient glows on the Moon were luminescence effects. The meteorite work, which was confirmed when lunar samples became available, showed that this was unlikely. Geake and Walker (1967) carried out an exhaustive examination of the luminescence spectra of meteorites under proton bombardment. The intensity and colour are governed by mineralogy, ranging from the very bright blue-red luminescence of aubrites (due to the enstatite) to the virtual non-luminescence of C chondrites. Other classes have about one-twentieth the brightness of aubrites and are various shades of green (due to feldspars), except for the ureilites which luminesce red (probably due to the diamonds).

A potentially more sensitive technique is thermoluminescence, in which the luminescence is stimulated by heating the specimen. Mineralogy is still the major factor governing the luminescence (see, for example, Sears and Mills 1974), and relative intensities and colours are much the same as in proton luminescence (Sears 1974). The thermal and radiation history of the specimen are also important, however, and thermoluminescence can be applied to determine quantities related to these factors. Because of its complexity, and the wealth of data available from other sources, little work has been done to develop these applications.

5.1.4. *Electrical and Magnetic Properties*

The electrical conductivity would be expected to be related in some fairly simple manner to the proportion of metal in the stone, and to some extent this is observed (Wood 1963). The low-metal C, LL and aubrite meteorites have conductivities near $10^{-8}\,(\Omega\,\text{cm})^{-1}$, while the comparatively metal-rich H chondrites have values of 10^{-3}–$10^{-4}\,(\Omega\,\text{cm})^{-1}$. L chondrites are intermediate in the amount of metal and in conductivity $(10^{-5}$–$10^{-6}\,(\Omega\,\text{cm})^{-1})$. Electrical conductivity is lowered considerably by the presence of water, and is four orders of magnitude lower in C1 chondrites than in CV3 and CO3 chondrites (Brecher *et al* 1975).

The natural remanent magnetism (NRM) in stony, stony-iron and iron meteorites is 0.5–20×10^{-3}, 5–70×10^{-3} and 10–300×10^{-3} emu g^{-1}, respectively (Guskova and Pochtarev 1969). The C chondrites examined by Brecher and Arrhenius (1974) ranged from 0.5×10^{-4} to 0.3 emu g^{-1}, and showed no correlation with petrological type. The main factor determining the NRM is the amount of ferromagnetic material present: metal and FeS in most chondrites, and magnetite in C chondrites (Herndon and Rowe 1974, Brecher and Ranganayaki 1975). However, other factors are also involved in iron meteorites. Guskova and Pochtarev (1969) found that the NRM intensity in iron meteorites varied with class; IAB, IIIAB and IVA had values of 30×10^{-3}, 65×10^{-3} and 120×10^{-3} emu g^{-1}, respectively. The groups have similar major-element chemistry, so the differences must be the result of different histories.

The aspect of magnetic studies which has received most attention is the estimation of the magnetic field strength in which the meteorite cooled (the 'palaeomagnetic field'). It has been shown that

$$\frac{M_0}{M_\mathrm{I}} = \frac{H}{H_\mathrm{I}}, \tag{5.1}$$

where M_0 is the NRM, M_I is the remanent magnetisation artificially induced by a magnetic field of strength H_I, and H is the palaeomagnetic field strength. A number of authors have used this relationship to measure the palaeomagnetic field strength. Values between 0.1 and 1.0 Oe have been obtained (Herndon and Rowe 1974), but there is some dispute as to the value of these determinations. Brecher and Arrhenius (1974) have argued that the results are meaningful only for the C chondrites, which have not been heated above the Curie points of their ferromagnetic materials. Herndon and Rowe

(1974) claim that C chondrites actually undergo reactions during the heating necessary to deduce M_I, so that even these values are unreliable. If, however, we accept the results at face value, then, when they cooled through the Curie point, the meteorites were in a magnetic field comparable with that of the Earth. The present-day interplanetary magnetic field strength is about 10^{-4} times this value, and is therefore insufficient to account for the meteoritic magnetism. It seems unlikely that the magnetic field in which meteorites cooled was generated by a dynamo effect, as in the case of the Earth, because meteorite parent bodies were not large enough (see § 3.5.2). The iron meteorites which are thought to be fragments of cores pose a particular paradox for this explanation. They would need to be molten to generate the field, and at the same time they would have to be solid to retain the magnetism (Herndon and Rowe 1974). A more feasible possibility is that the early interplanetary magnetic field was 10^4 times stronger than it is now: certain theories for the origin of the solar system include such a requirement (see, for example, Hoyle 1960). Meadows (1973) has suggested that the magnetism was induced by shock, in which case the above estimates of the magnetic field strength would be upper limits.

5.2. Radiogenic Properties

5.2.1. General Theory

The fundamental equation describing the rate of change in abundance, N_p, at time t of a parent nuclide undergoing radioactive decay, is

$$\frac{dN_p}{dt} = -\lambda N_p, \tag{5.2}$$

where the constant of proportionality, λ, is the decay constant. The initial and final abundance of the parent nuclide may be found by integration. The additional daughter product abundance will correspond to the difference. Therefore,

$$N_{d_o} = N_{p_o}[\exp(\lambda t) - 1] + N_{d_i}, \tag{5.3}$$

where the subscripts d, p, o and i denote daughter, parent,

126

observed and initial, respectively. The time t is now the period required for the nuclides to change their abundance from the initial to the final values. Isotope abundances are frequently expressed as ratios of a stable isotope, as these are more readily measured. A plot of N_{d_o} against N_{p_o} should therefore yield two quantities which are of significance (figure 5.2). These are, firstly, the slope of the line (the 'isochron') which is equal to $[\exp(\lambda t) - 1]$ and may be used to calculate the age of the meteorite; and secondly, the intercept, which equals the initial abundance of the daughter nuclide.

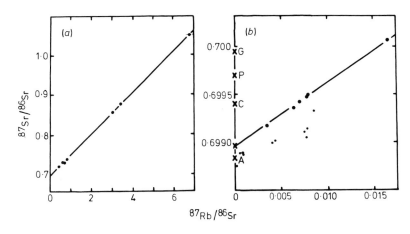

Figure 5.2. (*a*) Internal isochron for the Norton County aubrite. The slope of the line corresponds to an age of $4 \cdot 7 \times 10^9$ years (Bogard *et al* 1967). (*b*) Rb–Sr data for Rb-poor samples (note the scale change) allows more accurate extrapolation to initial conditions. The large dots refer to data obtained from six Ca-rich achondrites which define a whole-rock isochron equal to an age of $4 \cdot 39 \times 10^9$ years, and extrapolate to an initial $(^{87}Rb/^{86}Sr = 0)$ $^{87}Sr/^{86}Sr$ ratio (BABI). The crosses indicate the initial $^{87}Sr/^{86}Sr$ ratios for Guarena (G), Peace River (P), Colomera silicates (C) and Angra dos Reis (A). The small points are data from Ca-Al-rich aggregates and chondrules from the Allende meteorite (Gray *et al* 1973).

The observed parent and daughter abundances may be obtained in a variety of ways. They may be measured (i) in several fragments from the same meteorite; (ii) in minerals separated from the same meteorite, and (iii) in the bulk powder from several co-genetic meteorites. (In cases (i) and (ii) the line defined is called an 'internal isochron', whilst for

127

case (iii) it is called a 'whole-rock isochron'.) Alternatively, the 'model age' of a single meteorite can be calculated from the observed parent and daughter abundances in the bulk powder, using equation (5.3). This requires that an initial value for the parent nuclide is assumed. It is best determined from samples of low parent nuclide abundance, because little extrapolation is then required to the initial values. Usually, the eucrite and howardite initial value is used for Rb/Sr (figure 5.2), with Canyon Diablo troilite for methods involving Pb isotopes (Papanastassiou and Wasserburg 1969, Tatsumoto *et al* 1973).

5.2.2. The Rubidium–Strontium System

Whole-rock Rb–Sr ages and initial $^{87}Sr/^{86}Sr$ ratios are given in table 5.2. All groups appear to have become isolated from the solar nebula at about $4 \cdot 6 \times 10^9$ years ago. Ages and ratios determined from internal isochrons (for example, figure 5.2(*a*)) are listed in table 5.3. They lead to the same conclusion; that the chondritic meteorites and the aubrites formed $4 \cdot 6 \times 10^9$ years ago. The meteorites with ages clearly distinct from this value are those which have undergone some

Table 5.2. Whole-rock Rb–Sr ages and initial isotope values for the meteorite groups (Kaushal and Wetherill 1969, Papanastassiou and Wasserburg 1969).

Group	Number	Age (10^9 years)	$(^{87}Sr/^{86}Sr)_i$
E	11	$4 \cdot 55 \mp 0 \cdot 08$	$0 \cdot 6993 \mp 0 \cdot 001$
H	13	$4 \cdot 69 \mp 0 \cdot 14$	$0 \cdot 6983 \mp 0 \cdot 0024$
L	5	$4 \cdot 48 \mp 0 \cdot 14$	$0 \cdot 7008 \mp 0 \cdot 0012$
LL	11	$4 \cdot 56 \mp 0 \cdot 15$	$0 \cdot 7005 \mp 0 \cdot 0015$
C†	4	$4 \cdot 46 \mp 0 \cdot 35$	$0 \cdot 7007$
Eucrites	7	$4 \cdot 39 \mp 0 \cdot 35$	$0 \cdot 698990 \mp$ $0 \cdot 000047$

† Murthy and Compston (1965). Kaushal and Wetherill (1969) were unable to define an isochron using nine C chondrites, but stated that their results were consistent with their H group values above.

Table 5.3. Internal isochron measurements of Rb–Sr ages and initial $^{87}Sr/^{86}Sr$ ratios.

Meteorite	Group	Age (10^9 years)	$(^{87}Sr/^{86}Sr)_i$
Indarch	E4	4·56	0·7005
Guarena	E4	4·56 ∓ 0·08	0·69950 ∓ 0·00015
Peace River	L6	4·56 ∓ 0·03	0·69970 ∓ 0·00010
Krahenberg	LL5	4·70 ∓ 0·01	0·6989 ∓ 0·0005
Olivenza	LL5	4·63	0·6994
Norton County	Aubrite	4·70 ∓ 0·13	0·700 ∓ 0·002
Kapoeta	Howardite	3·6 − 3·9	
Nakhla	Anomalous achondrite	1·13 − 1·37	0·7023
Silicates in iron meteorites			
Colomera	IIE	4·61 ∓ 0·04	0·69940 ∓ 0·00004
Kodaikanal	IIE–An	3·8 ∓ 0·1	0·71 ∓ 0·02
Weekeroo Station	IIE	4·37 ∓ 0·2	0·703 ∓ 0·003

kind of melting process (§ 4.1). These are Kapoeta (howardite), Nakhla (anomalous achondrite), and the silicates from the Kodaikanal IIE–An iron.

Figure 5.2(*b*) illustrates some determinations of initial $^{87}Sr/^{86}Sr$ ratios. All measurements are from materials low in Rb—note the scale change from figure 5.2(*a*)—so that extrapolation to the initial $^{87}Sr/^{86}Sr$ ratio is minimised. The dots are six eucrites whose initial ratio is often termed BABI (basaltic achondrite best initial). An anomalous achondrite, Angra dos Reis, and the low-Rb inclusions from Allende, all have extremely low $^{87}Sr/^{86}Sr$ ratios, and any reasonable extrapolation to their initial values would result in a ratio much lower than BABI. Also shown are the initial values for Guarena (H6), Peace River (L6) and silicates from the Colomera IIE iron. The significance of different initial isotope ratios is not at all clear. One possibility is that the primordial nebula was not homogeneous in these isotopes. Another is that they reflect small differences in the time of accretion. The $^{87}Sr/^{86}Sr$ ratio in the primordial nebula, after the end of nucleosynthesis, increased at a rate of about 0·037% per million years because of [87]Rb decay. The difference in initial ratio between the lowest Allende point and BABI could result from Allende forming 10 million years or so before the eucrites. The same argument if applied to Guarena, Peace River and the Colomera silicates, would indicate that these meteorites did not form until up to 74 million years later, but other isotopes show that this is clearly not the case (§ 5.2.5). A third possibility is that the initial ratios of these meteorites were increased by some other mechanism, and it is the time at which this was operative that is being measured. In the chondrites, this mechanism may have been metamorphism, since both Guarena and Peace River are type 6. In the case of the iron meteorite, the melting event may have been responsible.

5.2.3. The Potassium–Argon System

A complication of the potassium–argon method is that ^{40}K decays not only to ^{40}Ar (by electron capture), but also to ^{40}Ca (by beta decay). A fraction $0 \cdot 110$ (the 'branching ratio') of the ^{40}K undergoes the former process. Virtually all the ^{40}Ar is radiogenic, so N_{d_i} may be set to zero, and equation (5.3) reduces to

$$^{40}\mathrm{Ar} = 0 \cdot 110\ ^{40}\mathrm{K}[\exp (\lambda t) - 1].\qquad (5.4)$$

Two methods have been used to measure the isotope abundances. The ^{40}Ar can be measured from gas release experiments, and the potassium can be determined chemically. Alternatively, the meteorite can be subjected to neutron irradiation, which induces the reaction ^{39}K(n, p)^{39}Ar (the ^{40}Ar–^{39}Ar method). The ^{39}Ar and ^{40}Ar can then be measured simultaneously by mass spectroscopy in the gas released on stepwise heating. This reaction is not 100% efficient, so that standards of known potassium content must be included in the irradiation (Turner 1969), or the ages obtained must be considered relative to an assumed age for one of them (Podosek 1971).

Histograms of the K–Ar ages (assuming an arbitrary 850 ppm for the potassium content) are shown in figure 5.3 (Zähringer 1968). H, E and LL chondrites have similar histograms. Most of these meteorites began to retain argon

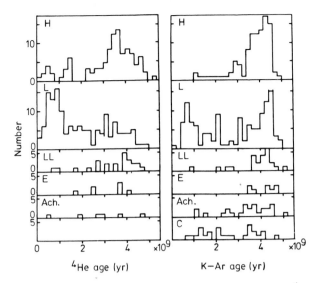

Figure 5.3. Formation ages calculated from radiogenic ^{4}He and ^{39}Ar contents. Note the large number of L chondrites with an apparent age of 0.5×10^9 years, and the difference between the Ar and He results for this group. The data for the C chondrites are not available because radiogenic He is swamped by trapped He (Zähringer 1968).

between 3.0 and 4.7×10^9 years ago, in agreement with their Rb–Sr formation ages. The L chondrites have a distinctly different histogram, and many meteorites in this class appear to have ages of about 0.5×10^9 years. We will return to this below. Four C1 chondrites have ages between 0.5 and 1.5×10^9 years, whilst the other C chondrite groups have K–Ar ages between 1 and 4×10^9 years (Heymann and Mazor 1968). These meteorites are extremely friable, and it may be that the low ages reflect the loss of Ar whilst in space.

Comparatively few meteorites have been examined by the ^{40}Ar–^{39}Ar method. Podosek (1971, 1972, 1973) has reported that three chondrites and two aubrites have ^{40}Ar–^{39}Ar ages of 4.6×10^9 years. Lafayette and Nakhla have ages similar to Nakhla's Rb–Sr age of 1.37×10^9 years, showing that the event which re-equilibrated the Sr isotopes also degassed the meteorites. Figure 5.4 shows Podosek's results for silicates from the El Taco iron meteorite. Each point represents the data obtained from the gases released at a certain temperature. They are therefore analogous to mineral separates, and

131

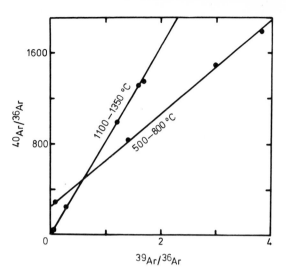

Figure 5.4. Composition of the gases released on stepwise heating of silicates from the El Taco iron meteorite after neutron irradiation. $^{40}Ar/^{36}Ar$ is a measure of the parent nuclide abundance (the actual parent nuclide is ^{40}K, which produces ^{40}Ar on neutron irradiation). $^{39}Ar/^{36}Ar$ is a measure of the daughter product abundance. The lines are therefore effectively internal isochrons. If the gases released at high temperatures (1100–1350 °C) are assumed to refer to an age of $4 \cdot 6 \times 10^9$ years, then the gases released at low temperatures (500–800 °C) correspond to an age of $3 \cdot 4 \times 10^9$ years (Podosek 1971).

effectively define an internal isochron. The gases released at 1130–1350 °C give an age equal to that of St Severin (assumed to be $4 \cdot 6 \times 10^9$ years), but the low-temperature gases give a slope corresponding to an age of $3 \cdot 4 \times 10^9$ years. It seems that the gas which was retained at the lower-retentivity sites (that is, that released at lower temperatures) was lost at this time. The $^{40}Ar-^{39}Ar$ method has also been used by Turner (1969) to determine the time of the last major outgassing event in eight chondrites. These all have K–Ar ages of $0 \cdot 5-1 \cdot 5 \times 10^9$ years when evaluated in the normal way. Stepwise heating experiments showed that they were outgassed $0 \cdot 3-0 \cdot 5 \times 10^9$ years ago. A further three chondrites examined by Turner had suffered no outgassing, and had ages of about $4 \cdot 5 \times 10^9$ years.

132

5.2.4. The Natural Decay Series

The natural decay series are utilised in two ways: by measurement of the ^{4}He produced by alpha decay (these are important steps in each series), and by measuring the abundance of lead isotopes, the decay products. ^{238}U, ^{235}U and ^{232}Th series produce eight, seven and six atoms of ^{4}He, respectively, so that equation (5.2) must be rewritten;

$$^4He_r = 8\,^{238}U[\exp(\lambda_{238}t)-1] + 7\,^{235}U[\exp(\lambda_{235}t)-1]$$
$$+ 6\,^{232}Th[\exp(\lambda_{232}t)-1]. \quad (5.5)$$

The radiogenic ^{4}He is calculated by subtracting the cosmogenic and trapped ^{4}He from the total helium. The uranium isotope abundances are usually averages from mass spectroscopic measurements. The major difficulty with the use of lead isotopes is that they diffuse rapidly, although it may be safely assumed that diffusion will affect all isotopes of the same nuclide equally, so their relative proportions should remain unaltered. Writing equation (5.3) for ^{235}U decay to ^{207}Pb, and ^{238}U decay to ^{206}Pb, and normalising to non-radiogenic ^{204}Pb, we have

$$\frac{^{207}Pb}{^{204}Pb} = \frac{^{206}Pb}{^{204}Pb}\frac{^{235}U[\exp(\lambda_{235}t)-1]}{^{238}U[\exp(\lambda_{238}t)-1]}. \quad (5.6)$$

The plot of ^{207}Pb/^{204}Pb against ^{206}Pb/^{204}Pb should define an isochron from which the age can be derived, provided that the ^{235}U/^{238}U ratio is known. Pb and U must be determined from the same sample, owing to their extreme heterogeneity in meteorites.

Zähringer's (1968) compilation of ^{4}He in chondrites are also presented as histograms in figure 5.3. They are based on a uniform U content of 11 ppb (parts per billion) which means that the error in any individual determination is very large; perhaps $\pm 40\%$. As with K–Ar ages, most members of the LL, H and E groups have ^{4}He ages between $3{\cdot}0$ and $4{\cdot}7 \times 10^9$ years. There is little or no $4{\cdot}6 \times 10^9$ year peak in the histogram of L group ^{4}He ages. This probably reflects more complete outgassing of the lighter ^{4}He than for ^{40}Ar. Heymann (1967) examined a number of shocked L chondrites, and found that two out of 16 fell in the 0.5×10^9 year peak. A literature survey suggested that between one-third and two-thirds of the class were involved. The most plausible explanation of the simultaneous shocking and outgassing of such a large proportion of a single class is to suppose that a collision occurred while they were part of the same parent body. On the other hand, Wänke (1966) and Öpik (1966) have suggested that it could be a consequence of diffusive gas loss. In this case, one would have to assume that shocked meteorites suffer diffusive gas loss more readily than unshocked meteorites.

Model ages, as well as ages based on whole-rock and internal isochrons, have been calculated from lead isotopes (Huey and Kohman 1973, Tatsumoto *et al* 1973, Tatsumoto and Unruh 1975). They give values of $4.5–4.6 \times 10^9$ years, in confirmation of the results based on other isotopes. The precision is sufficiently high for small apparent time differences to be measurable. Two eucrites (Nuevo Laredo and Sioux County) have ages which are smaller than Angra dos Reis by 27 million years. This compares with 14 million years suggested by the initial $^{87}Sr/^{86}Sr$ ratio. In both instances, an isotopically homogeneous nebula has to be assumed.

5.2.5. The Decay of Extinct Nuclides

The presence of extinct nuclide decay products was predicted by Brown (1947). They were discovered when the Richardton meteorite was found to contain excess ^{129}Xe, the decay product of ^{129}I (Reynolds 1960). ^{129}I must have been present in the meteorite when it accreted, in order for the ^{129}Xe to be retained. The isotope decays with a very short half-life (17 million years), so that accretion must have occurred within about 200 million years of the end of nucleosynthesis. However, complex models of nucleosynthesis are necessary to reconcile long half-life isotope measurements with this result, and estimates of the 'formation interval' are model-dependent (see, for example, Hohenberg 1969).

The decay of ^{129}I to ^{129}Xe occurred sufficiently rapidly for measurements of ^{129}Xe to provide information on the relative ages of formation of different meteorites. In the nebula, prior to accretion, the ratio of ^{129}I to stable ^{127}I changed due to decay. Upon accretion the ^{129}I decay product, ^{129}Xe, was retained and now provides a measure of ^{129}I. The isotope measurement procedure is the same as for the ^{40}Ar–^{39}Ar method. The meteorite specimen is irradiated with neutrons, and the reaction ^{127}I(n, $\gamma\beta$)^{128}I is induced. ^{128}I then undergoes beta decay to ^{128}Xe. ^{129}Xe and ^{128}Xe may then be measured by stepwise heating in a vacuum system linked to a mass spectrometer. Since

$$\frac{^{129}I}{^{127}I} = \frac{^{129}Xe}{^{128}Xe}\,\frac{^{128}Xe}{^{127}I}, \tag{5.7}$$

the ^{129}I/^{127}I ratio can be calculated from the measurements of ^{129}Xe/^{128}Xe, the second term on the right-hand side being known from calibration experiments. The time difference in accretion between two meteorites is then

$$dt = \tau\left[\lg\left(\frac{^{129}I}{^{127}I}\right)_1 - \lg\left(\frac{^{129}I}{^{127}I}\right)_2\right], \tag{5.8}$$

where τ is the mean life of ^{129}I. Results for 14 meteorites are shown in figure 5.5 (Wetherill 1975, Drozd and Podosek

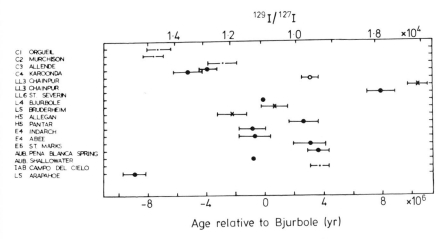

Figure 5.5. Formation intervals relative to Bjurbole estimated from the Xe decay products of now extinct ^{129}I. × Chondrules; ○ matrix; • other phases; ● bulk. See text for details of the method and the significance of the upper scale. The 'other phases' are magnetite from Orgueil and Murchison, Ca–Al-rich inclusions from Allende, and silicates from Campo del Cielo (Wetherill 1975, Drozd and Podosek 1976).

1976). All the meteorites studied, including the silicates from the Campo del Cielo IAB iron, began to retain [129]Xe within 20 million years of each other. This process appears to be independent of petrological type, whereas initial [87]Sr/[86]Sr ratios for the higher petrological types of chondrite were distinctly higher than for the lower types. It appears, therefore, that metamorphism was able to re-equilibrate the Sr isotopes without driving off the [129]Xe (Gray *et al* (1973). Lafayette and Petersburg (anomalous achondrite and howardite, respectively) do not have any excess [129]Xe, and lower limits for their formation intervals are 66 and 20 million years after the formation of St Severin (Podosek 1972).

The decay products of several other extinct nuclides have been sought. Recently, positive evidence for [244]Pu—which decays by fission and is discussed below—was reported, but attempts to discover the products of [107]Pb and [205]Pb have so far been unsuccessful.

5.2.6.   Fission and Charged-Particle Tracks

The fission-like xenon spectrum shown by the heavier isotopes [131]Xe, [132]Xe [134]Xe and [136]Xe in C chondrites and ordinary chondrites has been described in § 4.7.5. A different, fission-like spectrum is shown by the heavier isotopes of xenon in Ca-rich achondrites (Rowe and Kuroda 1965). Kuroda (1960) had previously speculated that these would be present as the fission products of extinct [244]Pu, which, like the other actinides, was expected to be comparatively rich in Ca-achondrites. Supporting evidence was rapidly acquired. Wasserburg *et al* (1969) found that the whitlockite ($Ca_3(PO_4)_2$), again actinide-rich, from St Severin (LL6) was rich in 'achondrite-type' xenon. Particularly pure xenon of this type was then found in the anomalously uranium-rich Angra dos Reis achondrite. Important evidence for the presence of [244]Pu in meteorites at some time was then obtained from charged-particle tracks (Canterlaube *et al* 1969, Wasserburg *et al* 1969). These are lines of radiation-damaged material produced by the passage through a crystal of heavy charged particles. For nuclei with atomic numbers greater than about 20, tracks may be observed under the optical microscope, after they have been enlarged by etching. A few studies have been made on the smaller tracks of lighter nuclei which require an electron microscope for their observation (Fleischer et al 1967, Maurette and Price 1975). The source of charged particles to produce these tracks is the spontaneous fission of unstable nuclei. Where tracks due to fission do not correlate with uranium abundance, there is strong evidence for ^{244}Pu. Recently, ^{244}Pu has been confirmed as the fissile nuclide by Alexander et al (1971), who examined the spectrum of the decay products of artificially produced ^{244}Pu, and found it to be identical with the achondrite-type xenon.

Theoretically, confirmation of the presence of ^{244}Pu decay products permits the determination of relative formation times in an analogous way to the I–Xe method, with the advantage that its longer half-life allows longer intervals to be measured. However, the Pu–Xe system presents a number of difficulties and, as yet, only a few results have been reported. Podosek (1972) reported a period of 146 million years for the Petersburg howardite, and one of greater than 350 million years for Lafayette. This compares with the lower limits of 20 and 66 million years from the I–Xe method.

5.2.7. The Aluminium–Magnesium System

One of the major problems in explaining the origin of meteorites is the source of heat which melted the achondrites, pallasites and certain iron meteorites. A possible heat source could have been ^{26}Al decay ($t_{1/2} = 7 \cdot 3 \times 10^5$ years). This would now be extinct; but, if there was enough to melt the meteorite parent bodies by the heat of its decay, it should also have produced detectable excess ^{26}Mg. This could not be found in ordinary chondrites (Schramm et al 1970), but it is present in the Ca–Al-rich Allende inclusions (Gray and Compston 1974, Lee and Papanastossiou 1974, Lee et al 1977). The ^{26}Mg excess is sometimes quite large, and reflects an amount of ^{26}Al initially present which would be enough to melt bodies larger than a few kilometres in diameter (see § 6.2.3).

5.3. Cosmogenic Properties

5.3.1. Introduction

The bombardment of a meteorite by cosmic rays has three effects: (i) it induces nuclear reactions in the elements of which the meteorite is composed; (ii) it leaves tracks of radiation-damaged material analogous to the tracks produced by fission (§ 5.2.7), and (iii) it can induce various luminescence phenomena. These three effects can be recognised as being due to cosmic ray bombardment because they show appreciable depth-dependence (figure 5.6).

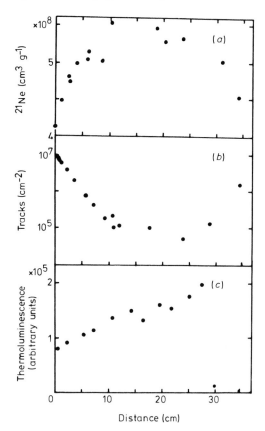

Figure 5.6. Variation with depth of three cosmogenic properties in the St Severin meteorite: (a) ^{21}Ne content; (b) charged-particle track density, and (c) thermoluminescence. The ^{21}Ne and the tracks, but not the thermoluminescence, were measured along the same core (Schultz et al 1973, Lal et al 1969, Lalou et al 1970).

The prime source of low-energy (1–100 eV) cosmic rays is the Sun, and the flux is directly related to the Sun's activity. The effects of these particles are confined to the outer centimetre or so of the meteorite, and for most purposes they can be ignored; ablation in the atmosphere has usually removed them. At higher energies (typically of the order of 1 GeV), the source is outside the solar system, and the cosmic rays are termed galactic, although their intensity is still affected by the Sun's activity. The particles are about 85% protons and 15% α particles, but at lower energies the proportion of heavy particles is higher (Kohman and Bender 1967). The primary flux is not necessarily solely responsible for the cosmogenic effects inside the meteorite. Spallation reactions produce a whole range of secondary particles which are added to the primary flux, and it is important that these should be taken into account.

The main aim of studies of cosmogenic properties has been to determine the interval during which meteorites have been exposed to cosmic radiation, this being referred to as the 'exposure age' or 'radiation age' of the meteorite. If the final abundance of the reaction products and their production rate in space are known, then the exposure age can be found by dividing the abundance by the production rate. The main unknown quantity in this relationship is the production rate for a given nuclide. This has to be estimated by theoretical or empirical methods. Readers not interested in the estimation of this quantity, or the details of cosmic ray interactions in meteorites, may proceed directly to § 5.3.4.

5.3.2. Production Rates

The production rate of a nuclide at depth d is given by

$$P(d) = N \int_0^\infty F(E, d)\sigma(E)\, dE, \qquad (5.9)$$

where N is the number of target atoms per unit area, $F(E, d)\, d(E)$ is the energy spectrum, and $\sigma(E)$ is the cross section for the reaction producing the nuclide. If a power law spectrum,

$$F(E)\, \mathrm{d}E = F_0 E^{-\alpha}\, \mathrm{d}E \qquad (5.10)$$

is assumed, then a solution to equation (5.9) is

$$P = k_1 (\Delta A)^{-k_2}, \qquad (5.11)$$

where k_1 is some constant, $k_2 = (3\alpha - 1)/2$, and ΔA is the difference in mass number between the target and product (Geiss *et al* 1962). This relationship was obtained empirically by Stauffer and Honda (1962). All else being equal, isotopes which are similar in mass number to their target isotopes have higher production rates than those with large ΔA. We can think of small ΔA isotopes as low-energy products (for example, ^{22}Na and ^{26}Al in stony meteorites, in which the targets are Mg and Si; ^{53}Mn in all meteorites, with Fe targets), and large ΔA isotopes as high-energy products (for example, ^{22}Na and ^{26}Al in irons, ^{3}H in all meteorites, with Mg, Si, Fe, O and S targets).

These relationships apply to incident radiation with a simple power law spectrum. Inside the meteorite, this no longer applies because secondary radiations have been added to the incident beam. We can allow for these in two ways. (i) We can adopt a different energy spectrum such as that of Reedy and Arnold (1972). These authors, using results from lunar studies, found that

$$F(E, d) = K(\gamma + E)^{-2.5}, \qquad (5.12)$$

where K is a normalisation constant (equivalent to F_0 in equation (5.10), and γ is a constant which allows for changes due to the addition of secondaries (the 'shape parameter'). Values of $\sigma(E)$ can be obtained from the literature. (ii) We can use a different reaction cross section such as that of Trivedi and Goel (1973). Instead of measuring $F(E, d)$—which is taken as the power law spectrum—$\sigma(E)$ is assumed to be a function of depth, $\sigma'(E)$. $\sigma'(E)$ must be determined in the laboratory by artificially irradiating a meteorite sample. It can then be calculated from

$$\sigma'(E) = \frac{Y}{\lambda I N n t}, \qquad (5.13)$$

where Y is the measured activity of the nuclide produced, λ is its decay constant, I is the intensity of the incident beam, N is Avagadro's number, n is the target atom number density, and t is the duration of the irradiation.

The results obtained by these two methods are in general agreement (figure 5.7). They show that the production of

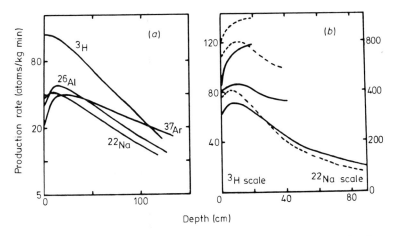

Figure 5.7. Estimates of the depth variation in the production rates of nuclides produced by cosmic ray bombardment. (*a*) Estimates of Reedy and Arnold (1972), which assume that the particle flux varies with depth, and employ reaction cross sections for thin targets; (*b*) results of Trivedi and Goel (1973), which assume a constant flux density, and employ reaction cross sections measured in thick targets as a function of depth. Broken curves, ^{3}H; full curves, [22]Na. Trivedi and Goel's results have been calculated for meteorites of a variety of sizes (20, 40 and 100 cm), while Reedy and Arnold's results apply to an infinitely large body.

high-energy nuclides decreases steadily with depth, but that low-energy products first increase with depth—due to the build-up of secondaries—and then decrease. We can therefore understand the depth variation in nuclide abundances shown in figure 5.6. Charged-particle tracks can be considered as a high-energy product, and thermoluminescence as a low-energy product.

k_2 is sometimes referred to as the 'irradiation hardness' and, although it is usually assumed to be constant, varies slightly with depth: it may be calculated from stable isotope pairs which show measurable depth variations (such as

^{21}Ne/^{22}Ne and ^{38}Ar/^{36}Ar; figure 5.6). For iron meteorites, k_2 varies between 2·1 for small meteorites and 3·1 for large meteorites, whilst for chondrites it lies between 1·9 and 2·2. These figures suggest that iron meteorites tend to be larger in space (Schultz and Hintenberger 1967, Nyquist *et al* 1973).

A second theoretical technique which is often used involves a relationship of the form

$$P = A \exp(\mu_a x) + B \exp(\mu_s x), \qquad (5.14)$$

where A, B, μ_a and μ_s are constants which involve nuclear parameters, target densities and other target characteristics, and flux intensity, and x is the depth into the specimen. The first term represents the contribution from primaries, whilst the second term allows for secondaries (Martin 1953, Ebert and Wänke 1957, Signer and Nier 1960). The constants are determined by fitting equation (5.14) to observed profiles. This approach is commonly used where laboratory data are scarce; for example, with spallogenic nuclides in stony meteorite metal (Nyquist *et al* 1973) and with thermoluminescence (Sears 1975a). Kohmann and Bender (1967) modified the treatment to allow for changes in irradiation harness: they summed two terms as in equation (5.14)—one to represent the high-, and one the low-energy particles—and allowed the relative amounts to vary with depth when fitting the curves.

5.3.3. Empirical Production Rates

The abundance of a rapidly decaying radioactive nuclide will soon reach a level where its production rate equals its rate of decay ('saturation'). This observation is the basis of all empirical production rate estimates. In certain cases, the production rate of a saturated isotope can be used to estimate the production rate of a stable isotope. Two isotopes are chosen which are isobaric, or isotopes of the same nuclide, so the energy dependence of their reaction cross sections will be very similar (for example, ^{3}H–^{3}He, ^{36}Cl–^{36}Ar, ^{38}Ar–^{39}Ar and ^{40}K–^{41}K). Since the two isotopes are measured in the same sample, they will have received the same particle flux. Then from equation (5.9)

$$\frac{P_1}{P_2} = \frac{\sigma_1}{\sigma_2}. \qquad (5.15)$$

If isotope 2 is radioactive and has reached saturation, we can substitute its decay rate, D_2, for its production rate. The production rate of the stable isotope 1 is then given by

$$P_1 = \frac{D_2 \sigma_1}{\sigma_2}. \qquad (5.16)$$

Usually, the radioactive nuclide decays so fast that its decay rate can only be measured in freshly fallen meteorites. Data are therefore very sparse. To study a reasonable number of meteorites it is necessary to make the sweeping assumption that the production rate in one sample is applicable to other samples, and even to other meteorites. It has been found that this is a particularly useful assumption for ^{26}Al, which is saturated in many meteorites, but not all. Its production rate in saturated meteorites (that is, its decay rate) can then be used to find the exposure age of meteorites in which the ^{26}Al has not reached saturation. Herzog and Anders (1971) have extended this method to estimate the production rates of the inert gas isotopes by dividing the inert gas abundance in a given meteorite by its ^{26}Al exposure age.

Production rates cannot be applied to meteorites of different classes unless allowance is made for their different target chemistries. This may be done empirically by fitting production rates (by the method of least squares) to an expression involving each expected major target element (Bogard and Cressy 1973, Mazor *et al* 1970, Cressy and Bogard 1976). For example, for ^{21}Ne we have

$$P(^{21}\text{Ne}) = 0{\cdot}0248[\text{Mg}] + 0{\cdot}00467[\text{Si}] + 0{\cdot}0033[\text{S}]$$

$$+ 0{\cdot}00093[\text{Ca}] + 0{\cdot}000239[\text{FeNi}], \qquad (5.17)$$

where the elements in square brackets represent abundances in weight per cent, and the production rate is measured in 10^{-8} cm^3 g^{-1} $(10^6$ yr$)^{-1}$ STP.

Production rates obtained by all these methods are given in table 5.4. It should be borne in mind that some studies use iron, and that others use silicate targets, and that an arbitrary

Table 5.4. Cosmogenic nuclide production rates determined by various methods†.

Nuclide	Semi-theoretical methods				Empirical methods	
	Equation (5.9) Method (i) (1)	(2)	Equation (5.9) Method (ii) (3)	Equation (5.14) (4)	Equation (5.16) (5)	^{26}Al method (6)
^{3}He			2·3–3·0		2·00	2·48
^{21}Ne			0·32		0·377	0·466
^{14}C	5·1/3·0	22		4·9/4·0		
^{26}Al	3·7/2·3	45		5·0/4·5		63
^{53}Mn	610/760	20		360/500		

† ^{3}He and ^{21}Ne production rates in $cm^3\, g^{-1}(10^6\, yr)^{-1}$ STP; others in dpm (disintegrations per minute)/kg.
[1] Arnold *et al* (1961). 25 cm iron meteorite at 1·3/13 cm depth.
[2] Reedy and Arnold (1972). Lunar surface at 50 cm depth.
[3] Trivedi and Goel (1969). Stony meteorite simulation, range observed in 80 cm target.
[4] Kohman and Bender (1967). 25 cm iron meteorite at 1·3/13 cm depth.
[5] Kirsten *et al* (1963). Stony meteorite.
[6] Heymann and Anders (1967) and Herzog and Anders (1971). Ordinary chondrite.

depth has been chosen for measurements concerned with depth profiles.

5.3.4. Exposure Age Results

The best exposure ages are obtained when the abundance of a stable isotope is measured in a meteorite sample to which equation (5.16) can be applied. In this case all the data are from a single sample. The drawback is that few such data exist. Some recent values obtained this way are given in table 5.5, and several earlier determinations have been listed by Anders (1963). The L chondrites have very low ages (2–10 million years), whereas two of the three H chondrites have values of 15–20 million years; but clearly no meaningful trends can be discerned from these meagre data.

The inert gas contents of a large number of meteorites are now known, and from these, the spallogenic gas abundances can be taken. For example, Zähringer's (1968) compilation of ^{3}He contents is presented in histogram form in figure 5.8.

Table 5.5. Exposure ages (10^6 years) calculated from radioactivity measurements (Bogard and Cressy 1973, Fireman and Goebel 1970).

Meteorite	Group	$^{22}Na/^{22}Ne$	$^{26}Al/^{21}Ne$
Allende	CV3		5·1
		4·9	4·0
Barwell	L5, 6	2·5	2·2
Bovedy			2·0
Kwaleni	H	24·3	19·6
Kako		5·0	3·2
Kiffa		15·3	12·3
Lost City	H5	6·1	4·8
Malakal	L	4·6	2·7
Nejo	L6	9·4	6·1
Tathlith	L	8·6	6·9
Ucera	H5	21	15
Wethersfield	L	6·2	4·1

Using the production rates described in the previous sections, isotope abundances have been converted to exposure ages. Such values neglect the variable effects of shielding. Judging from figure 5.7, this could introduce errors of up to 30% but, by using large numbers of specimens, the effect should cancel out. ^{21}Ne abundances give almost identical results (Kirsten *et al* 1963, Wasson 1974).

Exposure ages are considerably lower than formation ages (typically 10^7 years, compared with $4·6 \times 10^9$ years). The greatest value has been found for the Norton County aubrite (220 million years), whilst the smallest appears to be for the Farmington chondrite (20 000 years). The meteorites must therefore either have lost their cosmogenic gases, or have been shielded from cosmic rays for the major part of their existence. The former explanation is not tenable since, in many cases, radiogenic gases have been retained for $4·6 \times 10^9$ years. Furthermore, in many meteorites ^{26}Al has not reached saturation, and this occurs after about 1 million years of exposure to cosmic radiation. Clearly, stony meteorites have not always had the same dimensions. The question is whether they have been continuously eroded, or have undergone one or two major impacts.

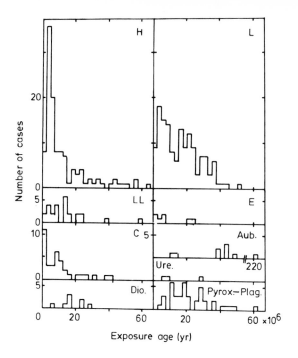

Figure 5.8. Duration of exposure to cosmic radiation ('exposure' or 'radiation' age) for stony meteorites, calculated by dividing the abundance of cosmogenic nuclides by their estimated production rates. The abbreviations Aub., Ure., Dio. and Pyrox.–Plag. refer to aubrites, urielites, diogenites and Ca-rich achondrites, respectively (Zähringer 1968, Ganapathy and Anders 1969, Eberhardt *et al* 1965a, Herzog and Cressy 1976, Mazor *et al* 1970).

The second important feature of the exposure age distributions are the peaks in the histograms. These indicate that a large number of specimens in a class became exposed to cosmic radiation at the same time; so, presumably, they were parts of a larger body which fragmented. The 4 million year H group cluster is well established. Clusters in the L group histogram are disputed (Tanenbaum 1967), and there is probably a continual spread from about 100 million years downwards, with greater numbers at lower values. A similar trend is likely for the E and LL chondrites. The C chondrites show a strong correlation between exposure age and petrological type; C1 and C2 chondrite ages are below 15 million years, whereas the other C chondrites go to higher values. This

146

probably reflects the friability of the lower groups. Data are inevitably very sparse for the achondrite groups. The eucrites and howardites show an identical pattern and range in exposure age up to about 60 million years. Unlike the chondrites, they do not show a large number of small ages. Ganapathy and Anders (1969) claim that there are 11, 19 and 22 million year clusters in these classes. Six of the nine aubrites examined by Eberhardt *et al* (1965a) plot in a cluster near 40 million years, older than most stony meteorites, whilst all the diogenites are around 20 million years. Herzog and Cressy (1976) have claimed to have resolved this into 14 and 24 million year clusters.

The most comprehensive study for iron meteorites is that of Voshage (1967), whose results, with those of Wänke (1966), are shown in figure 5.9. The exposure ages of iron meteorites are considerably greater than those of stony meteorites—going up to 10^9 years—but are still significantly less than their formation ages. The smallest value is for Pitts (IAB) and the greatest is for Deep Springs (an anomalous

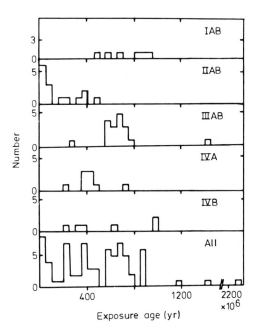

Figure 5.9. Exposure ages of iron meteorites calculated from $^{40}K–^{41}K$ and ^{21}Ne, using $^{4}He/^{21}Ne$ to make a depth correction (Voshage 1967, Wänke 1966).

iron) (4 and 2300 million years, respectively). The IIIAB and IVA groups have exposure ages which tend to cluster around 700 million years and 400 million years, whilst IAB and IVB exposure ages show a wide spread. The IIAB irons also show a spread in their exposure ages, but all are significantly lower than IIIAB.

We will now look more deeply into the question of why the exposure ages are less than the formation ages. The clusters of exposure ages in certain groups suggest that there have been a number of discrete major break-ups of the parent body in which the meteorite class was located. Subsequent fragmentation, by thermal stressing or additional impacts, may result in different ages for these individual meteorites. Different exposure ages for fragments of the same meteorite probably reflect such a multiple break-up. This being so, it seems that the aubrites, with their large exposure ages, must have been in an orbit which kept them away from collisions; whereas the H chondrites must have travelled through a collision-prone region and recently suffered a major impact (4 million years ago). A number of additional arguments have been advanced in favour of the major collision hypothesis over that of continual erosion. These are based on the stony meteorite size distribution, the mechanical strength of irons and stones, the absence of ^{36}Ar (which would be abundant in small objects because of cosmogenic n,γ reactions), and observed dust densities from zodiacal light measurements (see Kirsten and Schaeffer (1971) for references), but most convincing are the cosmogenic tract studies which will be discussed in the next section.

5.3.5. Cosmogenic Charged-Particle Tracks

Most track studies have concentrated on tracks produced by the heavy iron group nuclides. These are present in the primary radiation in very small amounts and they are attenuated rapidly with depth. Fleischer *et al* (1967a) give the following expression for the trap density D, at time t and depth R_0:

$$\frac{dD}{dt} = 2\left(\frac{dN}{dE}\right)\left(\frac{dE}{dR}\right)_{R_0} dR(z)F \exp(-\psi R_0) \cos \alpha \sin \theta,$$

$$(5.18)$$

where (dN/dE) is the energy spectrum (see equation (5.10)), $(dE/dR)_{R_0}$ is the rate of energy loss estimated from simulation experiments, $dR(z)$ is the track length, F is the fragmentation factor, ψ is the interaction probability and α and θ are geometrical terms. The results of this calculation may be compared with measurements such as those in figure 5.6. It is then possible to make an accurate estimate of the amount of material ablated from the meteorite during its passage through the atmosphere. From this, estimates of pre-atmospheric mass can be made, and fragments of the same fall can be re-assembled (see, for example, Cantelaube *et al* 1969, Price *et al* 1967). Fleischer *et al* (1967b) calculated the burial depths of 36 chondrites, and numerous subsequent values have been reported.

Equation (5.18) predicts the rate of build-up of the track number with time. If the meteorite were being eroded by interplanetary dust at an equal rate, there would be no currently observable cosmogenic tracks. From the maximum track density on the observed surface of the Patwar meteorite, Price *et al* (1967) calculated that the erosion in space was less than 10^{-7} cm yr^{-1}. For iron meteorites Fisher (1967) calculated 10^{-8} cm yr^{-1}. The rate of ablation required to produce the observed exposure age distributions can be calculated by writing the erosion rate in centimetres per year as

$$R = \frac{L}{T},$$ (5.19)

where L is the mean absorption length (say 1 m). We may take the exposure age, T, to be 20 million years for stones and 500 million years for irons. The required ablation rates are then 10^{-5} cm yr^{-1} for stones, and 5×10^{-8} cm yr^{-1} for irons. The ablation rate required, therefore, is greater than the maximum consistent with the track data.

Track data also provide important information on the origin of gas-rich meteorites (Lal and Rajan 1969, Wilkening *et al* 1971). Large track numbers and steep gradients have been observed in individual grains from Kapoeta and in the dark portions of meteorites with a light–dark structure (§ 4.7.2). The random location of these grains, and the track densities and gradients in them, suggests that these meteorites were once part of the surface soil (regolith) of a parent body. Similar findings have been reported for lunar samples (Lal 1972).

5.4. Non-Cosmogenic and Non-Radiogenic Isotope Studies

5.4.1. Introduction

In general, the relative proportions of the various isotopes of an element are much the same in terrestrial samples, lunar samples and meteorites. Most exceptions can be ascribed to radiogenic and cosmogenic processes. This has led to a widespread assumption that the solar nebula, from which these bodies formed, was originally well mixed and isotopically homogeneous. With the increased sensitivity of equipment in recent years, however, a number of elements have been identified with isotope anomalies which cannot be explained by these mechanisms. As well as this, some very small isotope differences, which can be interpreted as being due to extremely small formation time differences, may simply reflect isotopic inhomogeneity in the primitive solar nebula (Cameron 1974, Huey and Kohman 1973, Drozd and Podosek 1976).

Several isotope anomalies, which cannot be ascribed to cosmogenic or radiogenic processes, have already been discussed; for example, the trapped inert gas isotopes (§ 4.7.1), and the carbon isotopes in C chondrites (§ 4.8.2). There is insufficient space to discuss the isotope anomalies in Hg (Reed and Jovanovic 1969), H (Boato 1954), S (Lewis and Krouse 1969) and Cs (Rosman and DeLaeter 1976), but we will discuss the important work on oxygen.

5.4.2. Oxygen

The $^{16}O/^{18}O$ ratio varies between the meteorite classes to an extent that would enable it to be used as a means of classification. In ordinary chondrites, the ratio increases systematically from H to L or LL, from 3·7 to 4·3 per mil (‰) (which is defined in an analogous way to $\delta^{13}C$ in § 4.8.2, but using mean ocean water as a standard). When the values for different groups are similar, a common or related origin is

150

suggested. This seems to be the case with the E chondrites and the aubrites, and with the diogenites, eucrites, howardites and mesosiderites. On the other hand, Shergotty (which resembles the eucrites and howardites in composition), has distinctly different oxygen isotope ratios and is probably unrelated (Taylor *et al* 1965, Reuter *et al* 1965). In the ordinary chondrites the difference in the $^{16}O/^{18}O$ ratio between pairs of minerals from the same meteorite depends on petrological type. This suggests re-equilibration during metamorphism. Laboratory calibration experiments can then be used to determine the equilibration temperature (Onuma *et al* 1972; § 3.2.3). In a similar way, oxygen isotopes in the bulk sample have been used to determine the temperature at which the bulk meteorite equilibrated with the nebula (that is, to measure its accretion temperature). The values obtained were all below 500 K. However, the discovery of oxygen isotope anomalies has effectively changed the calibration curve so that, for CV chondrites at least, these values are probably inaccurate (Onuma *et al* 1974).

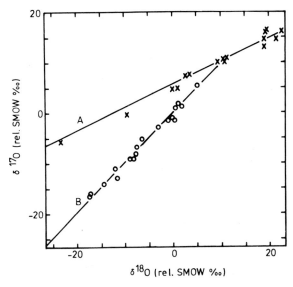

Figure 5.10. Oxygen isotope anomaly in Ca–Al-rich aggregates from the Allende and similar meteorites. The isotopes are expressed in parts per mil (‰) relative to standard mean ocean water (SMOW). The crosses which define curve A refer to terrestrial and lunar samples; the circles which define curve B apply to the Allende aggregates (Clayton *et al* 1973).

The oxygen isotope anomalies were discovered by Clayton *et al* (1973) in Ca–Al-rich aggregates in the Allende (CV3) meteorite. Line A in figure 5.10 results from chemical effects analogous to those used to determine metamorphic temperatures above. The terrestrial and lunar rocks, together with most meteorites, lie on this line. Inclusions from Allende lie along line B, which is a mixing line between ordinary oxygen and pure ^{16}O. Additional ^{16}O is therefore present from a source other than that which provided the ordinary oxygen. Clayton *et al* (1973) have suggested that it was pre-solar interstellar dust. Clayton (1977) has even suggested that the Ca–Al aggregates condensed, not in the solar nebula, but in the expanding shell of a supernova. This idea has been pursued by Lattimer *et al* (1977).

6.   History and Origin

6.1.   The Formation and Early History of Meteorites

6.1.1.   Summary of Compositional Data

The major meteorite classes, the chondrites, appear to have been formed by a condensation and accretion process in the primordial solar nebula. Details of the mechanism are disputed: the one that has been worked out in the greatest detail incorporates equilibrium condensation. The oxidation–reduction conditions, and therefore the mineralogy, of the classes can be readily explained for C and ordinary chondrites. C chondrites formed below 400 K, and ordinary chondrites formed between 500 and 600 K. The trends in oxidation state within the ordinary chondrites can be explained by accretion in the order H, L and LL. E chondrites, however, require high pressures and temperatures to obtain the level of reduction in this class, but it is then difficult to explain their contents of sulphur and other volatile elements. It seems that either non-equilibrium condensation, condensation from a gas of non-solar composition, or a complex accretion process are required for this class.

The compositional differences between the chondrite classes, and the patterns of trace elements within them, may be explained by *ad hoc* processes superimposed on this basic condensation model. Four processes have been identified. (i) Differences in the Mg/Si ratio are assumed, in all groups, to be due to the removal of high-temperature (Mg-rich) condensates. On the other hand, the Ca/Si and Al/Si ratios in CV chondrites require the addition of a high-temperature condensate. (ii) Differences in the Fe/Si ratio are probably caused by the separation of iron and silicates. The process responsible appears to be different in the E chondrites, where the S/Fe ratio varies with the Fe/Si ratio. (iii) Moderately

153

volatile elements have been lost, possibly during the chondrule-forming episode, or by gas loss during their condensation. (iv) Highly volatile elements are depleted by factors of up to 1000 in the higher petrological types. The interpretation of this depletion is keenly contested: the major proposals require either metamorphism or partial condensation during accretion. Heating experiments favour the partial condensation model for the ordinary chondrites, but are unable to offer constructive conclusions for the other classes. Cooling rates probably also favour the partial condensation model.

In iron meteorites, several elements (Ni, Ga and Ge)—which appear to have been little affected by post-formation events—provide data on the formation process. Group IVB represents the first metal to condense from the nebula. The other groups most probably accreted between 600 and 700 K, but over a wide pressure range. Their relative abundances of sulphur and carbon may also have been determined during condensation and accretion. Most of the other elements in iron meteorites have been fractionated within the groups, although much less so in IAB. The mechanism responsible was either fractional crystallisation or partial melting. Cooling rates favour partial melting.

With the exception of the aubrites, the achondrite groups seem to represent the products of melting chondrite-like material. The Ca-rich achondrites (eucrites and howardites) could be produced by fractional crystallisation or partial melting, whilst the ureilites appear to be the residual solid left after partial melting. The aubrites share many compositional trends with the E chondrites, and they have similar oxygen isotope ratios: the two groups are probably co-genetic. The aubrites have no metal and chondrules, and may represent E chondrites which accreted early, that is, before metal and chondrule formation.

There are two small classes of stony-iron meteorites. The pallasites represent an olivine layer which has been intruded by metal: such a situation could occur at an interface between the metal inclusion, or core, and the surrounding silicates. Mesosiderite silicates resemble Ca-rich achondrites, and have similar oxygen isotope ratios. However, they have a distinctly different Ca/Al ratio and are probably not co-genetic. The metal, in fact, seems to be related to the IIE iron meteorites. The mesosiderites appear to be a mechanical mixture of materials similar to these different classes.

6.1.2. Summary of Chronological Data

The condensation and accretion process seems to have occurred very quickly $4 \cdot 6 \times 10^9$ years ago, and soon after the end of nucleosynthesis (§ 5.2). Most chondrites then underwent metamorphism, but the date at which this occurred is poorly known. The initial $^{87}Sr/^{86}Sr$ ratio in Guarena suggests that it was 80 million years after the formation of the Ca-rich achondrites, but this is not compatible with the I–Xe results, unless it is assumed that metamorphism can equilibrate the strontium isotopes and not lose the xenon (§ 5.2.5).

Xe will almost certainly have been lost during the melting events which produced the iron meteorites and achondrites. The I–Xe and Pu–Xe methods therefore provide a date for the melting. In two achondrites (Petersburg and Lafayette), this occurred 146 and more than 350 million years, respectively, after the formation of the chondrites. Silicates from the Kodaikanal (IIE–An) iron have a Rb–Sr age of $3 \cdot 8 \times 10^9$ years. The melting processes which produced these meteorites therefore occured some time after accretion, and continued for perhaps 10^9 years (Gray *et al* 1973).

6.2. Meteorite Parent Bodies

6.2.1. Evidence for Meteorite Parent Bodies

The concept of meteorite parent bodies (as well as their size and internal structure) is one of the most important features of current meteorite research. When Chladni originally postulated that meteorites were extraterrestrial, he imagined that they existed as the same metre-sized chunks that entered the atmosphere. We now know that this is incorrect, and we have to assume that the meteorites belonged to larger parent bodies. We may mention three pieces of evidence for this. (i) Exposure ages for meteorites are less than their formation ages; typically 50 million years for stones and 1000 million years for irons, compared with 4600 million years. The difference cannot be explained by micrometeorite erosion, but must be due to fragmentation by multiple collisions (§ 5.3.4). (ii) Most of the L chondrites suffered severe outgassing 500 million years ago. The simplest explanation is that this occurred when they were part of the same object. The outgassing may even have been caused by the disruption of that object. Subsequent secondary collisions have reduced the fragments even more, and produced the lower exposure ages. In many meteorite classes (such as IIIAB and IVA irons, H chondrites, aubrites and diogenites) a large number, or perhaps all, of their members were exposed to cosmic radiation at the same time, presumably by fragmentation of their mutual parent body. (iii) Meteorites can be divided into classes on the basis of their composition. This suggests that each class came from a unique set of conditions and probably, therefore, a separate parent body. If they came from different regions of a reservoir showing the complete spectrum of composition, we would not expect compositional 'gaps'.

6.2.2. The Number of Meteorite Parent Bodies

Anders (1971) has argued that the number of parent bodies from which meteorites have come is very few. The chondrites are sufficiently similar to have originated from just three or four parent bodies. The iron meteorite groups require about twelve. In fact, the IIIAB irons (with exposure ages of 600–700 million years) may have been involved in the break-up of the L chondrite parent body 500 million years ago, either as the other colliding object, or because they were all part of the same body. (A systematic difference of 100 million years between the two sets of results cannot be ruled out because the methods used have been different.) Like many of the L chondrites, most of the IIIAB irons have been shocked (Jaeger and Lipschutz 1967). It is possible to argue that parent bodies are also required for the six achondrite classes, about twelve anomalous chondrites and achondrites, and more than fifty anomalous iron meteorites. However, these may equally be regarded as random freaks, or the products of melting, or mixing, of the major classes. In terms of the mass, or numbers

156

of meteorites falling on the Earth, only the chondrite parent bodies are significant.

6.2.3. The Size of Meteorite Parent Bodies and Their Heat Source

The suggestion that meteorites come from lunar-sized bodies has been made several times. These arguments were based on systems which were thought to require high pressures: the absence of the Widmanstätten pattern in high-nickel meteorites, and the presence of cohenite and diamonds. Alternative explanations appear to be valid for each of these features (§§ 3.5.2 and 3.5.3). On the other hand, Mason (1962) has argued that the presence of tridymite in meso-siderites and E chondrites suggests that these meteorites have not suffered pressures in excess of 3 kbar. At the moment, therefore, there seems to be no compelling reason to believe that meteorites have been subjected to long-term high pressures.

The only quantitative estimates for meteorite parent body sizes are based on cooling rates (§ 3.5.2). These suggest bodies a few hundred kilometres in diameter, but this figure is specu-lative in that it assumes an abrupt removal of the heat source, and that the body was initially above 600 °C. The latter is probably a reasonable assumption for all meteorites. The metamorphism in most chondrites shows that they have been heated to at least this temperature (§ 3.2.3); whilst the Wid-manstätten pattern in most iron meteorites shows that they have been heated into the γ field of the phase diagram (§ 3.5.2; figure 3.9). The assumption that the heat was removed abruptly is less certain. It seems more resonable that the heat would subside slowly, in which case these estimates would be upper limits.

There is no agreement on the heat source responsible for this initially high temperature. Long half-life radioactivity will suffice only if the body is extremely large, and the presence of tridymite argues against this. The only short half-life radioac-tivity which is a serious possibility is ^{26}Al. Upon accretion, the amount of ^{26}Al in the Allende Ca–Al-rich aggregates would have been sufficient to melt all but the smallest objects, but by then it may already have been extinct. However, it is not clear how typical the aggregates are. Other possibilities which have been discussed are gravitational contraction (Urey 1962, Ostic 1965), or induced electric fields during the Sun's T Tauri phase (Sonnet *et al* 1970). Full quantitative treatments have either yet to be made, or give pessimistic results.

6.2.4. The Structure of Meteorite Parent Bodies

The cooling rate estimates for iron meteorites seem particularly relevant to the question of the structure of the meteorite parent bodies. According to Goldstein and co-workers (§ 3.5.2), all groups except IAB and IIIB (the Ni-rich half of IIIAB) show a wide range of cooling rates. This implies a wide range of burial depths, so the meteorites in groups IIAB, IIIA and IVA would have been isolated bodies in a silicate matrix, rather than cores. There is ample trace element and mineralogical evidence that IAB groups were never in the core of a parent body. This leaves only IIIB groups as possible candidates for core material, but a common burial depth for isolated bodies is equally plausible. Although these cooling rate data have been disputed by Willis and Wasson (1977), it is also difficult to explain why all the groups (except IVB) contain sulphides: these would soon separate from the metal if the iron were completely melted in a core. At the moment the evidence would seem to support the belief that no meteorite parent body had a metallic core.

A few stony meteorites have been melted. A significant proportion of these, like lunar soil, are gas-rich. Some also contain charged-particle track evidence for being part of a surface regolith at some time. The most plausible way to produce brecciation, melts and partial melts from regolith material is by meteoritic bombardment of a 'planetary' surface. A hypothetical, and highly speculative, composite meteorite parent body for all the meteorite classes is shown in figure 6.1. Localised melting, due to meteorite impact, we imagine, has produced eucrites and howardites. The diogenites may then form as a by-product (cumulates). Beneath this level, only partial melting occurs. The low-melting-point, high-density metal and metal–sulphide eutectics have flowed

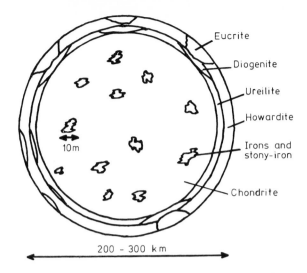

Figure 6.1. Hypothetical composite parent body for many of the meteorite classes. E chondrites, CB chondrites, CV chondrites, and all but one of the iron meteorite classes require separate parent bodies.

downwards; whereas the low-melting-point, low-density feldspar and similar minerals have flowed upwards into the howardite/eucrite regions. We may further imagine that the metal has been partially melted over a period of time, and flowed to various depths, forming 1–10 m bodies. The bulk of our imaginary 200–300 km main body is chondritic. We require separate main bodies for the E, ordinary (perhaps even one for each of the H, L and LL classes), CB and CV chondrites. On, or near, the surface of at least one of them the eucrites, howardites, ureilites and diogenites formed; whilst inside them, or within a number of similar bodies, the iron meteorites formed.

6.3. Data Bearing on Postulated Sources

6.3.1. *Nebular Location of the Condensation Process*

The degree of reduction is greater for higher pressures and temperatures in the nebula. Assuming adiabatic conditions, the pressure and temperature of the nebula should decrease with increasing distance from the Sun. It seems generally

159

agreed, therefore, that the distance at which accretion occurred increased along the series E, H, L, LL and C; with the E–H and LL–C gaps much greater than the others. However, there is no agreement over details. Larimer and Anders (1967) place the E chondrites on the inner fringes of the asteroid belt, and the C chondrites no nearer than the outer fringes of the belt. If the E chondrites formed from a gas of solar composition, then they formed very close to the Sun (where the pressure was $1-10^{-2}$ atm; § 4.2), perhaps within the orbit of Mercury. The accretion conditions of the iron meteorite groups require a 10^4-fold variation in nebular pressure (probably from 10^{-4} to 10^{-8} atm), corresponding to a 10-fold difference in the distance from the Sun.

The condensation calculations have been fitted into a reasonable pressure–temperature curve (adiabat) by Lewis (1972) to explain the density, and predict the composition, of the terrestrial planets. When fitted to this 'empirical' adiabat, ordinary chondrites formed at 1 AU, and C chondrites formed in the asteroid belt. On the other hand, when fitted into Cameron and Pine's (1973) theoretical adiabat, the ordinary chondrites could be formed in the asteroid belt.

6.3.2. The Moon and Planets

Numerous dynamical arguments have been advanced against meteorites originating on the Moon. Some of these will be apparent below. However, the most compelling evidence is provided by the returned lunar samples, and there is now universal disbelief in the Moon as a source of meteorites. Chemically and mineralogically, the lunar samples resemble only the eucrites and howardites (Gast 1972), but even in this case there are substantial lead and oxygen isotope differences (Tatsumoto et al 1971, Taylor and Epstein 1970).

The planets have also been universally rejected as a potential source of meteorites. The escape velocity that would be necessary to eject them would require a highly energetic explosion. This would turn them into material resembling the glassy spherules found around meteorite craters or, at best, tektite-like glasses (§§ 2.3.3 and 2.4).

6.3.3. Orbital Data

Although several hundred meteorites fall on the Earth each year, we have sufficient data to derive reliable orbits for only three of them. These are the Lost City, Oklahoma (H5), Pribram, Czechoslovakia (H6) and Innisfree, Alberta (L5, 6) meteorites which fell into areas covered by camera networks set up to detect bright meteors (figure 6.2). The information so provided is severely limited. Earth-crossing orbits are extremely short-lived, and the meteorites could not have been formed in them. However, alternative data on meteorite orbits exist. Eye-witness observations, although too unreliable for velocity measurements, are of value for radiant observations and fall times. A second kind of relevant data is provided by the exposure ages: these may be characteristic of certain orbits, since the collision which exposed the meteorite to cosmic radiation may also have placed it on its final orbit.

Also shown in figure 6.2 is the orbit of one of the 23 Apollo asteroids. These are objects with Earth-crossing orbits. Wetherill (1976) considers them to form a single family

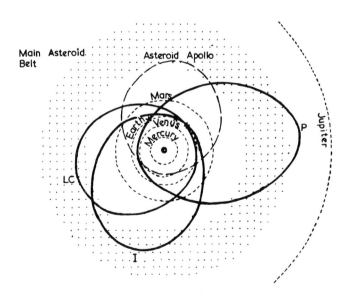

Figure 6.2. The solar system, showing the orbits of the planets out to Jupiter, the main asteroid belt, and the three known meteorite orbits: P, Pribram; LC, Lost City, and I, Innisfree (Fireman and Spannagel 1971, Halliday *et al* 1977).

with the 22 Amor asteroids, which have perihelia between 1·02 and 1·30 AU. The Apollo and Amor asteroids have orbits which closely resemble the known meteorite orbits, and almost certainly represent the last stage of a meteorite's journey to Earth. The orbit of the Farmington meteorite has been calculated from eye-witness reports (Levin *et al* 1976). Although far inferior to those based on photographs, it also resembled several Apollo orbits. This is particularly significant in view of Farmington's extremely short exposure age (20 000 years), so that the meteorite and its parent body were probably still in the same orbit when it landed.

Precise computations of orbits soon become unreliable when pursued for times in excess of 1000 years. Statistical methods are therefore used, the most popular being Arnold's (1965) Monte Carlo method. This is a numerical model which assumes random encounters with planets, and that the probability of a meteorite coming within a distance R of the centre of a planet is

$$P = \frac{R^2 u}{2u^{-2} \sin i a_0 a^2 (1-e)^{1/2} |\cot \alpha|}, \tag{6.1}$$

where u is the relative velocity, a, e and i are the semi-major axis, eccentricity and inclination of the orbit of the meteorite, and α is the Sun–meteorite–planet angle (Öpik 1966). The calculations show that the Earth-crossing orbits are stable only for periods of 10^6 years, after which time and meteorite becomes Venus-crossing (about 10^5 years), and is soon ejected from the solar system by the influence of Jupiter. The Apollo asteroids and the meteorites were therefore clearly not formed in these orbits. The Apollo, Amor and meteorite orbits closely resemble the orbits of the short-period comets, and they pass through the asteroid belt. The most likely sources for meteorites, the author concludes, are the short-period comets or the main-belt asteroids.

6.3.4. Comets

Most of the major meteor showers are known to be associated with comets. An association between meteorites and comets is therefore intuitively attractive. The main proponents of such a connection are Öpik (1966) and Wetherill (1968, 1971). An observation that may be relevant is the preference for afternoon rather than morning falls; that is, a preference for radiants with small elongations (figure 6.3; Farrington 1915, Mason 1962). Wetherill (1968) performed Monte Carlo calculations with initial orbits resembling those of the Moon, an

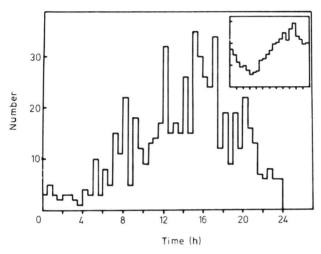

Figure 6.3. Histogram showing the fall of all stony meteorites in the *Catalogue of Meteorites* (Hey 1966). Inset is the predicted histogram for objects originating from a short-period comet (Wetherill 1968).

Apollo asteroid, and a short-period comet with an aphelion near Jupiter's orbit. The last, which is also equivalent to a Trojan asteroid, agreed reasonably well (see inset in figure 6.3). The same calculations yield predicted lifetimes which can be obtained by combining probabilities of encounters. If the meteorite was put on its final orbit by a collision, then lifetimes are equivalent to exposure ages. Histograms of these predicted lifetimes, calculated for a short-period comet, bear a striking similarity to the exposure age histograms for chondrites, especially for the H chondrites.

The major evidence in favour of a cometary origin for stony meteorites, is thus that it predicts appropriate exposure ages (Wetherill 1968, Öpik 1966). Arguments against this origin are, primarily, the following. (i) The non-volatile cores of comets are too small for them to be meteorite parent

163

bodies. For example, the core of Encke's comet is only 10 km in diameter. (ii) It is not clear that complex chemical fractionations, such as those observed in chondrites, could occur in the source region of the comets. This is thought to be well outside the orbit of Pluto. (iii) The conventional picture of the nucleus of a comet is of a highly friable object which is incapable of surviving atmospheric entry. However, data on all these points are scarce. More recently, Anders (1975b) has argued that since gas-rich meteorites belonging to all classes have solar flare tracks and solar-type implanted gases, they could not have formed more than 8 AU from the Sun; that is, they could not have formed where the comets are thought to have originated.

6.3.5. Asteroids

The major theme in dynamical studies of the origins of meteorites has been how asteroids, and so potential meteorite parent bodies, can be extracted from the main asteroid belt. An obvious way to change the orbit of an asteroid is by collision with another, assuming that their density in space is high enough. However, Anders (1964) showed that sufficiently energetic collisions would destroy the objects. This even applies to a two-stage mechanism involving transfer to a Mars-crossing orbit followed by perturbation by Mars (Arnold 1964). Two methods have been proposed which do not require energetic collisions, but involve resonances with Jupiter. The orbits of the Lost City and Pribram meteorites, when extrapolated back in time, seem to be consistent with this. It is doubtful, however, that the efficiency would be sufficient to produce all the meteorites, or that these mechanisms would produce exposure ages less than $4 \cdot 6 \times 10^9$ years (Zimmerman and Wetherill 1973, Williams 1973, Wetherill 1976, Lowrey 1971). Dynamically, therefore, it seems very difficult to understand how main-belt asteroids could be parent bodies for all but a few meteorites. In any case, spectral reflectivity data, taken at face value, argue against an origin for the ordinary chondrites among the main-belt asteroids (§ 5.1.3), but this is not the case for the C chondrites, or for certain achondrites.

Anders (1964) has proposed that the 34 asteroids already in Mars-crossing orbits are the parent bodies of meteorites. Monte Carlo calculations for objects initially in these orbits predict lifetimes of 10^8–10^9 years. Although these are too high for stony meteorites, they seem to be appropriate to iron meteorites. Mars-crossing asteroids also appear to be of the correct size and number. The reality of the break-up of the L chondrite parent body 500 million years ago is important in this connection (§ 5.2.4), because it renders more likely an origin for this class among the Mars asteroids.

6.4. Summary

There now seems to be general agreement that the meteorites formed in the primordial solar nebula at the same time as the planets, $4 \cdot 6 \times 10^9$ years ago, by a similar condensation process. More highly reduced meteorites were nearest the Sun and the most oxidised furthest out. During the condensation and accretion process, a certain amount of material became molten droplets which were quenched to form the chondrules. This period of formation lasted a very short time (about 10^7 years), and occurred soon after the elements had been synthesised (a few $\times 10^8$ years), but the latter period is uncertain because of difficulties in reconciling long and short half-life radioactive element abundances. High-precision isotope abundance determinations suggest that either very small differences in formation time (about 10^6 years) can be measured, or that the primordial nebula was inhomogeneous in certain isotope ratios (Pb, Sr, O, Mg and Ne). There is also the possibility that solid material from outside the solar system was incorporated into the meteorites.

Most meteorites show signs of having been heated to temperatures high enough to change their structure and composition (up to about 1000 °C). A few meteorites (most achondrites and many irons) have been partially, or completely, melted. The heat source for this process is unknown, and the possibility that it was part of the formation process cannot be ruled out. Many of them have a light–dark structure and are rich in gases implanted by the solar wind.

Several gas-rich meteorites are also rich in tracks of radiation-damaged material.

Two pieces of evidence suggest that the number of parent bodies from which meteorites have originated are very few; perhaps less than a dozen. (i) The majority of meteorites are remarkably uniform in composition and structure. The few which are highly diverse in these properties may usually be interpreted as alteration products. (ii) The abundance of isotopes produced by cosmic ray bombardment shows that meteorites were released from larger bodies by a number of discrete collisions. For compositional and dynamical reasons, the Moon and the planets can be eliminated as potential meteorite parent bodies. The Apollo asteroids are the immediate parent objects, but their orbits can only exist for a short time. They must therefore be the products of another source: short-period comets or asteroids. It is extremely difficult to develop theoretical models to remove asteroids from the main asteroid belt, although inefficient methods exist. It is easier for Mars-crossing asteroids; although even these require time periods comparable with the exposure ages of iron meteorites (about 1000 million years), or to a postulated 500 million year event for stones. The reality of this latter event is questionable. In general, the stony meteorite exposure ages, especially the very short C chondrite ages, are comparable with the time expected to convert a short-period comet orbit to an Earth-crossing orbit. Complex element abundance patterns in meteorites can be explained in terms of physical and chemical conditions in the inner solar system, but too little is known about conditions in the regions where comets formed to say whether they could have occurred there.

References

Ahrens L H 1964 *Geochim. Cosmochim. Acta* **28** 411
—— 1965 *Geochim. Cosmochim. Acta* **29** 801
Ahrens L H, Von Michaelis H, Erlank E J and Willis J P 1969 *Meteorite Research* ed P M Millman (Dordrecht, Holland: Reidel) p 166
Alexander E C, Lewis R S, Reynolds J H and Michel M C 1971 *Science* **172** 837
Anders E 1963 *The Solar System IV: The Moon, Meteorites and Comets* ed B M Middlehurst and G Kuiper (Chicago: Chicago University Press) p 420
—— 1964 *Space Sci. Rev.* **3** 583
—— 1971 *Physical Studies of the Minor Planets* ed T Gehrels *NASA Spec. Publ.* SP 267 p 429
—— 1975a *Meteoritics* **10** 283
—— 1975b *Icarus* **24** 363
—— 1977 *Earth Planet. Sci. Lett.* **36** 14
Anders E, Hayatsu R and Studier M H 1973 *Science* **182** 781
Anders E and Heymann D 1969 *Science* **164** 821
Anders E, Higuchi H, Ganapathy R and Morgan J W 1976 *Geochim. Cosmochim. Acta* **40** 1131
Angra 1977 *Earth Planet. Sci. Lett.* **35** 271
Arnold J R 1964 *Isotope and Cosmic Chemistry* ed H Craig *et al* (Amsterdam: North-Holland) p 347
—— 1965 *Astrophys. J.* **141** 1536
Arnold J R, Honda M and Lal D 1961 *J. Geophys. Res.* **66** 3519
Arrhenius G and Alfvén H 1971 *Earth Planet. Sci. Lett.* **10** 253
Baedecker P A and Wasson J T 1975 *Geochim. Cosmochim. Acta* **39** 735
Baker G 1958 *Am. J. Sci.* **256** 369
Baldwin B and Shaeffer Y 1971 *J. Geophys. Res.* **76** 4653
Bandursky E L and Nagy B 1976 *Geochim. Cosmochim. Acta* **40** 1397
Barnes V E 1967 *International Dictionary of Geophysics* vol 2, ed S K Runcorn (Oxford: Pergamon) p 1507
—— 1971 *Chemie Erde* **30** 13
Barnes V E and Barnes M A 1973 *Tektites* (Stroudsburg, Pennsylvania: Dowden, Hutchinson and Ross)
Baudin 1798 *Phil. Mag.* **2** 225
Bauer C A 1947a *Phys. Rev.* **72** 354
—— 1947b *Phys. Rev.* **74** 225

Bauer C A 1947c *Phys. Rev.* **74** 501

Beals C S, Innes M J S and Rottenberg J A 1963 *The Solar System IV: The Moon, Meteorites and Comets* ed B M Middlehurst and G Kuiper (Chicago: Chicago University Press) p 253

Begemann F and Vilcsek E 1969 *Meteorite Research* ed P M Millman (Dordrecht, Holland: Reidel) p 355

Begemann F and Wänke H 1969 *Meteorite Research* ed R M Millman (Dordrecht, Holland: Reidel) p 363

Berzelius J J 1834 *Ann. Phys.* **33** 113

Binns R A 1967 *Nature* **213** 1111

Binns R A, Cleverley W H, McCall G J H, Reed S J B and Scoon J H 1977 *Meteoritics* **12** 179

Binz C M, Ikramuddin M, Rey P and Lipschutz M E 1976 *Geochim. Cosmochim. Acta* **40** 59

Bjork R L 1961 *J. Geophys. Res.* **66** 3379

Black D C 1972a *Geochim. Cosmochim. Acta* **36** 347

—— 1972b *Geochim. Cosmochim. Acta* **36** 377

Black D C and Pepin R O 1969 *Earth Planet. Sci. Lett.* **6** 395

Blander M 1971 *Geochim. Cosmochim. Acta* **35** 61

—— 1975 *Geochim. Cosmochim. Acta* **39** 1315

Blander M and Abdel-Gawad M 1969 *Geochim. Cosmochim. Acta* **33** 706

Blander M and Katz J L 1967 *Geochim. Cosmochim. Acta* **31** 1025

Boato G 1954 *Geochim. Cosmochim. Acta* **6** 209

Boctor N Z, Meyer H O A and Kullerud G 1976 *Earth Planet. Sci. Lett.* **32** 69

Boeckl R 1972 *Nature* **236** 25

Bogard D D, Burnett D S, Eberhardt P and Wasserburg G J 1967 *Earth Planet. Sci. Lett.* **3** 179

Bogard D D and Cressy P J 1973 *Geochim. Cosmochim. Acta* **37** 527

Brandurski E L and Nagy B 1976 *Geochim. Cosmochim. Acta* **40** 1397

Brecher A and Arrhenius G 1974 *J. Geophys. Res.* **79** 2081

Brecher A, Briggs P L and Simmons G 1975 *Earth Planet. Sci. Lett.* **28** 37

Brecher A and Ranganayaki R P 1975 *Earth Planet. Sci. Lett.* **25** 57

Brentnall W D and Axon H J 1962 *J. Iron Steel Inst.* **200** 947

Brett R 1966 *Science* **153** 60

Brett R and Higgins G T 1969 *Geochim. Cosmochim. Acta* **33** 1473

Brezina A 1904 *Proc. Am. Phil. Soc.* **43** 211

Briggs M H and Mamikunian G 1963 *Space Sci. Rev.* **1** 647

Brown H 1947 *Phys. Rev.* **72** 348

—— 1960a *J. Geophys. Res.* **65** 1679

—— 1960b *J. Geophys. Res.* **66** 1316

Brown H, Kullerud G and Nichiporuk W 1953 *A Bibliography on Meteorites* (Chicago: Chicago University Press)

Buchwald V F 1974 *Handbook of Iron Meteorites* (Berkeley: University of California Press)

Bunch T E, Keil K and Olsen E 1970 *Contrib. Miner. Petrol.* **25** 297

Bunch T E and Reid A M 1975 *Meteoritics* **10** 303

Buseck P R and Goldstein J I 1969 *Bull. Geol. Soc. Am.* **80** 2141
Cameron A G W 1973 *Icarus* **18** 377
—— 1974 *Nature* **246** 30
Cameron A G W and Pine M R 1973 *Icarus* **18** 377
Cantalaube Y, Maurette M and Pellas P 1967 *Radioactive Dating and Low-level Counting* (Vienna: IAEA) p 215
Cantalaube Y, Pellas P, Nordemann D and Tobailem J 1969 *Meteorite Research* ed P M Millman (Dordrecht, Holland: Reidel) p 705
Carr M H 1970 *Geochim. Cosmochim. Acta* **34** 689
Case D R, Laul J C, Pelly I Z, Wechter M A, Schmidt-Bleek F and Lipschutz M E 1972 *Geochim. Cosmochim. Acta* **36** 19
Cassidy W A, Olsen E and Yanai K 1977 *Meteoritics* **12** 190
Chang C and Wänke H 1969 *Meteorite Research* ed P M Millman (Dordrecht, Holland: Reidel) p 397
Chapman C R 1976 *Geochim. Cosmochim. Acta* **40** 701
Chapman C R and Salisbury J W 1973 *Icarus* **19** 507
Chladni E F F 1794 *Über den Ursprung der von Pallas gerfundenen und anderer ihr ahnlicher Eisenmasse und über einig damir in Berkindung stebende Naturerscheinungen* (Riga: Harttnoch)
—— 1819 *Über Feure-Meteore, und über die mit denselben herangefallen Massen* (Vienna: Heuber)
Clayton D D 1977 *Earth Planet. Sci. Lett.* **36** 381
Clayton R N, Grossman L and Mayeda T K 1973 *Science* **182** 485
Cressy P J and Bogard D D 1976 *Geochim. Cosmochim. Acta* **40** 749
Dietz R S 1977 *Meteoritics* **12** 145
Doan A S and Goldstein J I 1969 *Meteorite Research* ed P M Millman (Dordrecht, Holland: Reidel) p 763
Dodd R T 1969a *Min. Mag.* **37** 230
—— 1969b *Geochim. Cosmochim. Acta* **33** 161
—— 1976 *Earth Planet. Sci. Lett.* **30** 381
Dodd R T, Grover J E and Brown G E 1975 *Geochim. Cosmochim. Acta* **39** 1585
Dohnanyi J S 1971 *Physical Studies of the Minor Planets* ed T Gehrels *NASA Spec. Publ.* SP 267 p 263
Drozd R J and Podosek F A 1976 *Earth Planet. Sci. Lett.* **31** 15
Duke M B and Silver L T 1967 *Geochim. Cosmochim. Acta* **31** 1637
Eberhardt P, Eugster O and Geiss J 1965a *J. Geophys. Res.* **70** 4427
Eberhardt P, Geiss J and Grögler N 1965b *Tschermaks Miner. Petrogr. Mitt.* **10** 535
Ebert K H and Wänke H 1957 *Z. Naturf.* **A12** 766
Egan W G, Ververka J, Noland M and Hidgeman T 1973 *Icarus* **19** 358
Ehmann W D, Baedecker P A and McKown D M 1970 *Geochim. Cosmochim. Acta* **3** 493
Eugster O, Eberhardt P and Geiss J 1967 *Earth Planet. Sci. Lett.* **3** 249
Fanale F P and Cannon W A 1972 *Geochim. Cosmochim. Acta* **36** 319
Farrington O C 1915 *Meteorites* (Chicago: Chicago University Press)
Fireman E L and Goebel R 1970 *J. Geophys. Res.* **65** 3035

Fireman E L and Spannagel F 1971 *Chemie Erde* **30** 83

Fisher D E 1967 *Radioactive Dating and Low-level Counting* (Vienna: IAEA) p 269

Fleischer R L, Price P B, Walker R M and Maurette M 1967a *J. Geophys. Res.* **72** 331

Fleischer R L, Price P B, Walker R M, Maurette M and Morgan G 1967b *J. Geophys. Res.* **72** 355

Florensky K P 1976 *NASA Tech. Transl.* TTF-765

Florensky K P, Short N, Winzer S R and Fredriksson K 1977 *Meteoritics* **12** 227

Folinsbee R E and Bayrock L A 1964 *R. Astron. Soc. Canada J.* **58** 109

Folinsbee R E, Bayrock L A, Cumming G L and Smith D G W 1969 *R. Astron. Soc. Canada J.* **63** 61

Fougeroux A D , Cadet L C and Lavoisier A 1772 *Obs. Phys. L'Hist. Nat., Arts (J. Physique)* **2** 251

French B M 1968 *Shock Metamorphism of Natural Materials* ed B M French and N M Short (Baltimore: Mono) p 475

Fricker P E, Goldstein J I and Summers A L 1970 *Geochim. Cosmochim. Acta* **34** 475

Ganapathy R and Anders E 1969 *Geochim. Cosmochim. Acta* **33** 775

Gast P W 1972 *The Moon* **5** 121

Gast P W and Hubbard N J 1970 *Earth Planet. Sci. Lett.* **10** 94

Gault D E, Quaide W L and Oberbeck V R 1968 *Shock Metamorphism of Natural Materials* ed B M French and N M Short (Baltimore: Mono) p 87

Geake J E and Walker G 1967 *Proc. R. Soc.* **296** 337

Geiss J, Oeschger H and Schwarz U 1962 *Space Sci. Rev.* **1** 197

Gentner W, Lippolt H J and Schaeffer O A 1963 *Geochim. Cosmochim. Acta* **27** 191

—— 1973 *Earth Planet. Sci. Lett.* **20** 204

Gerling E K and Levski L K 1956 *Dokl. Akad. Sci. USSR* **110** 750

Goldberg E, Uchijama A and Brown H 1951 *Geochim. Cosmochim. Acta* **2** 1

Goldschmidt V M 1929 *Proc. R. Inst. Gt Br.* **26** 73

Goldstein J I and Ogilvie R E 1965 *Geochim. Cosmochim. Acta* **29** 893

Goldstein J I and Short J M 1967a *Geochim. Cosmochim. Acta* **31** 1001

—— 1967b *Geochim. Cosmochim. Acta* **31** 1733

Gordon R B 1970 *J. Geophys. Res.* **75** 439

Graham A L, Easton A J and Hutchison R 1977 *Min. Mag.* **41** 201

Gray C M and Compston W 1974 *Nature* **251** 495

Gray C M, Papanastassiou D A and Wasserburg G J 1973 *Icarus* **20** 213

Grossman L 1972 *Geochim. Cosmochim. Acta* **36** 597

—— 1973 *Geochim. Cosmochim. Acta* **37** 119

Guskova E G and Pochtarev V I 1969 *Meteorite Research* ed P M Millman (Dordrecht, Holland: Reidel) p 633

Haidinger W 1869 *Rep. Br. Assoc. Adv. Sci.* **39** 300

Halliday I, Blackwell A T and Simmons G 1978 *J. R. Astron. Soc. Canada* **72** 15

Haverö 1972 *Meteoritics* **7** 515

Hawkins G S 1960 *Astron. J.* **65** 318

Hayatsu R 1966 *Science* **153**

Hayatsu R, Matsuaka S, Scott R G and Studier M H 1977 *Geochim. Cosmochim. Acta* **41** 1325

Hayatsu R, Studier M H, Oda A, Fuse K and Anders E 1968 *Geochim. Cosmochim. Acta* **32** 175

Hayes J M 1967 *Geochim. Cosmochim. Acta* **31** 1395

Heide F 1964 *Meteorites* transl E Anders and E DuFresnes (Chicago: Chicago University Press)

Henderson E P and Perry S H 1954 *Geochim. Cosmochim. Acta* **6** 221

Herbig C H 1974 *Am. Sci.* **62** 200

Herndon J M and Rowe M W 1974 *Meteoritics* **9** 289

Herndon J M and Suess H E 1976 *Geochim. Cosmochim. Acta* **40** 395

Herzog G F and Anders E 1971 *Geochim. Cosmochim. Acta* **35** 605

Herzog G F and Cressy P J 1976 *Geochim. Cosmochim. Acta* **41** 127

Hey M 1966 *Catalogue of Meteorites* (London: British Museum, Natural History)

Heymann D 1967 *Icarus* **6** 189

—— 1971 *Handbook of Elemental Abundances in Meteorites* ed B Mason (London: Gordon and Breach) p 29

Heymann D and Anders E 1967 *Geochim. Cosmochim. Acta* **31** 1793

Heymann D, Lipschutz M E, Nielsen B and Anders E 1966 *J. Geophys. Res.* **71** 619

Heymann D and Mazor E 1968 *Geochim. Cosmochim. Acta* **32** 1

Higgins W 1818 *Phil. Mag.* **51** 355

Higuchi H, Morgan J W, Ganapathy R and Anders E 1976 *Geochim. Cosmochim. Acta* **40** 1563

Hintenberger H, Vilcsek E and Wänke H 1965 *Z. Naturf.* **A20** 939

Hohenberg C M 1969 *Science* **166** 212

Howard E C 1802 *Phil. Trans.* **51** 355

Hoyle F 1960 *Q. J. R. Astron. Soc.* **1** 28

Huey J M and Kohman T P 1973 *J. Geophys. Res.* **78** 3227

von Humboldt A 1849 *Cosmos: A Sketch of a Physical Description of the Universe* vol 1 (London: Bohn)

Hutchison R, Bevan A W R and Hall J M 1977 *Appendix to the Catalogue of Meteorites* (London: British Museum, Natural History)

Ikramuddin M, Binz C M and Lipschutz M E 1976 *Geochim. Cosmochim. Acta* **40** 133

—— 1977 *Geochim. Cosmochim. Acta* **41** 393

Ikramuddin M and Lipschutz M E 1976 *Geochim. Cosmochim. Acta* **39** 363

Ivanova I A, Lebedinets V N, Maksakov B I and Portnyagin Y I 1968 *Geochim. Int.* **190**

Jaeger R R and Lipschutz M E 1967 *Geochim. Cosmochim. Acta* **31** 1811

Jain A V and Lipschutz M E 1968 *Nature* **220** 139

Jarosewich E 1967 *Geochim. Cosmochim. Acta* **31** 1103

Jarosewich E and Mason B 1969 *Geochim. Cosmochim. Acta* **33** 411

Johnson A A and Remo J L 1974 *J. Geophys. Res.* **79** 1142

Johnson T V and Fanale F P 1973 *J. Geophys. Res.* **78** 8507

Jungclaus G, Cronin J R, Moore C B and Yuen G U 1976 *Nature* **261** 126

Kaula W M 1968 *An Introduction to Planetary Physics—The Terrestrial Planets* (New York: Wiley)

Kaushal S H and Wetherill G W 1969 *J. Geophys. Res.* **74** 2717

Keays R R, Ganapathy R and Anders E 1971 *Geochim. Cosmochim. Acta* **35** 337

Keil K 1968 *J. Geophys. Res.* **73** 6945

Keil K and Fredriksson K 1964 *J. Geophys. Res.* **69** 3487

Kelly W R and Larimer J W 1977 *Geochim. Cosmochim. Acta* **41** 93

Kengott A 1869 *Phil. Mag.* **38** 424

Kenna 1976 *Geochim. Cosmochim. Acta* **40** 1427 (series of papers)

King E A , Butler J C and Carman M F 1972 *Proc. 3rd Lunar Sci. Conf.* (Cambridge, Mass: MIT Press) p 673

Kirsten T, Krankowsky D and Zahringer J 1963 *Geochim. Cosmochim. Acta* **27** 13

Kirsten T A and Schaeffer O A 1971 *Elementary Particles* ed L C L Yuan (New York: Academic Press) p 75

Kohman T P and Bender M L 1967 *High Energy Nuclear Reactions in Astrophysics* ed W A Shen (New York: Benjamin) p 169

Krinov E L 1960 *Principles of Meteoritics* (New York: Pergamon)

—— 1966 *Giant Meteorites* (New York: Pergamon)

Kuroda P K 1960 *Nature* **187** 36

—— 1971 *Nature* **230** 40

Kuroda P K and Manuel O K 1970 *Nature* **227** 1113

Lal D 1972 *Space Sci. Rev.* **14** 3

Lal D, Lorin C, Pellas P, Rajan R S and Tamhane A S 1969 *Meteorite Research* ed P M Millman (Dordrecht, Holland: Reidel) p 275

Lal D and Rajan R S 1969 *Nature* **223** 269

Lalou C, Nordmann D and Labeyrie J 1970 *C. R. Acad. Sci. Paris* **270** 2401

Lancet M S and Anders E 1973 *Geochim. Cosmochim. Acta* **37** 1371

Larimer J W 1968 *Geochim. Cosmochim. Acta* **32** 965

—— 1973 *Geochim. Cosmochim. Acta* **37** 1603

Larimer J W and Anders E 1967 *Geochim. Cosmochim. Acta* **31** 1239

—— 1971 *Geochim. Cosmochim. Acta* **34** 367

Larimer J W and Buseck P R 1974 *Geochim. Cosmochim. Acta* **38** 471

Lattimer J M, Schramm D N and Grossman L 1977 *Nature* **269** 116

Laul J C, Keays R R, Ganapathy R, Anders E and Morgan J W 1972 *Geochim. Cosmochim. Acta* **36** 329

Lawless J G, Zeitman B, Perieira W E, Summons K E and Duffield A M 1974 *Nature* **251** 40

Lee T and Papanastassiou D A 1974 *J. Geophys. Res. Lett.* **1** 227
Lee T, Papanastassiou D A and Wasserburg G J 1977 *Astrophys. J.* **211** L107
Levin B J, Simovenko A N and Anders E 1976 *Icarus* **28** 307
Lewis C F and Moore C B 1971 *Meteoritics* **6** 195
Lewis J S 1972 *Earth Planet. Sci. Lett.* **15** 286
Lewis J S and Krouse H R 1969 *Earth Planet. Sci. Lett.* **5** 425
Lipschutz M E 1964 *Science* **143** 1431
Lockyer J N 1890 *Meteoritic Hypothesis* (London: Macmillan)
Lovering J F 1957 *Geochim. Cosmochim. Acta* **12** 238
Lovering J F, Nichiporuk W, Chodos A and Brown H 1957 *Geochim. Cosmochim. Acta* **11** 263
Lovering J F, Parry L G and Jaeger J C 1960 *Geochim. Cosmochim. Acta* **19** 156
Lowrey B E 1971 *J. Geophys. Res.* **76** 4084
McCarthy T S and Ahrens L H 1971 *Earth Planet. Sci. Lett.* **11** 35
McCarthy T S, Ahrens L H and Erlank A J 1972 *Earth Planet. Sci. Lett.* **15** 86
McCarthy T S, Erlank A J and Willis J P 1973 *Earth Planet. Sci. Lett.* **18** 433
McCord T B, Adams J B and Johnson T V 1970 *Science* **168** 1445
McCrosky R E and Ceplecha Z 1969 *Meteorite Research* ed P M Millman (Dordrecht, Holland: Reidel) p 600
McCrosky R E, Posen A, Schwartz G and Shao C-Y 1971 *J. Geophys. Res.* **76** 4090
Marcus H L and Hackett L H 1974 *Meteoritics* **9** 371
Marcus H L and Palmberg P W 1971 *J. Geophys. Res.* **76** 2095
Marti K 1967 *Earth Planet. Sci. Lett.* **3** 243
Martin G R 1953 *Geochim. Cosmochim. Acta* **3** 288
Maskelyne N S 1870 *Phil. Trans.* **160** 189
Mason B 1962 *Meteorites* (New York: Wiley)
—— 1963a *Am. Mus. Novit.* 2155
—— 1963b *Am. Mus. Novit.* 2163
—— 1967a *Geochim. Cosmochim. Acta* **31** 107
—— 1967b *Am. Sci.* **55** 429
—— (ed) 1971 *Handbook of Elemental Abundances in Meteorites* (London: Gordon and Breach)
—— 1972 *Meteoritics* **7** 309
Mason B and Jarosewich E 1971 *Meteoritics* **6** 241
—— 1973 *Min. Mag.* **39** 204
Mason B, Nelen J A, Muir P and Taylor S R 1976 *Meteoritics* **11** 21
Maurette M and Price P B 1975 *Science* **187** 121
Mazor E, Heymann D and Anders E 1970 *Geochim. Cosmochim. Acta* **34** 781
Meadows A J 1973 *Nature* **237** 274
Meinschein W G 1963 *Space Sci. Rev.* **2** 653
Miller S L 1953 *Science* **117** 528

Millman P M (ed) 1969 *Meteorite Research* (Dordrecht, Holland: Reidel) p 541

Moore C B, Lewis C F and Nava D 1969 *Meteorite Research* ed P M Millman (Dordrecht, Holland: Reidel) p 738

Moren A E and Goldstein J I 1977 *Meteoritics* **12** 318

Müller O, Baedecker P A and Wasson J T 1971 *Geochim. Cosmochim. Acta* **35** 1121

Murthy R V and Compston W 1965 *J. Geophys. Res.* **70** 5297

Nagy B 1975 *Carbonaceous Meteorites* (Amsterdam: Elsevier)

Nagy B, Meinschein W G and Hennessy D J 1961 *Ann. NY Acad. Sci.* **93** 25

Nininger H H 1936 *Am. J. Sci.* **32** 1

—— 1956 *Arizona's Meteorite Crater: Present—Past—Future* (Sedona: American Meteorite Museum)

Noddack I and Noddack W 1930 *Naturwissenschaften* **18** 757

Nyquist L, Funk H, Schultz L and Signer P 1973 *Geochim. Cosmochim. Acta* **37** 1655

O'Keefe J A 1976 *Tektites and Their Origins* (Amsterdam: Elsevier)

Olsen E J, Bunch T E and Keil K 1977 *Meteoritics* **12** 109

Onuma N, Clayton R N and Mayeda T K 1972 *Geochim. Cosmochim. Acta* **36** 157

—— 1974 *Geochim. Cosmochim. Acta* **38** 189

Öpik E J 1966 *Adv. Astron. Astrophys.* **4** 302

—— 1968 *Ir. Astron. J.* **8** 185

Oro J and Skewes H B 1965 *Nature* **207** 1042

Osborne T W , Smith R H and Schmitt R A 1973 *Geochim. Cosmochim. Acta* **37** 1909

Osborne T W, Warren R G, Smith R H, Wakita H, Zellmev D L and Schmitt R A 1974 *Geochim. Cosmochim. Acta* **38** 1359

Ostic R G 1965 *Mon. Not. R. Astron. Soc.* **131** 191

Pallas P S 1776 *Reisen durch verschiedene Provinzes des Russischen* vol 3 (St Petersburg: Die Kaiserlischen Akademie der Wissenschaften) p 411

Paneth F A, Gehlen H and Guenther P L 1928 *Z. Electrochem.* **34** 645

Paneth F A, Reasbeck P and Mayne K I 1952 *Geochim. Cosmochim. Acta* **2** 300

Papanastassiou D A and Wasserburg G J 1969 *Earth Planet. Sci. Lett.* **5** 361

Pellas P and Storzer D 1976 *The Interrelated Origin of Comets* ed A H Delsemme (Toledo: University of Toledo Press)

Perry S H 1944 *Metallography of Meteoritic Iron, USNM Bull.* 184

Podosek F A 1971 *Geochim. Cosmochim. Acta* **35** 157

—— 1972 *Geochim. Cosmochim. Acta* **36** 755

—— 1973 *Earth Planet. Sci. Lett.* **19** 135

Powell B N 1969 *Geochim. Cosmochim. Acta* **33** 789

—— 1971 *Geochim. Cosmochim. Acta* **35** 5

Price P B, Rajan R S and Tamhane A S 1967 *J. Geophys. Res.* **72** 1377

Prior G T 1916 *Min. Mag.* **18** 26
—— 1918 *Min. Mag.* **18** 151
—— 1920 *Min. Mag.* **19** 51
Prinz M, Hlava P H and Keil K 1974 *Meteoritics* **9** 393
Randich E and Goldstein J I 1975 *Meteoritics* **10** 479
Reed G W and Jovanovic S 1969 *J. Inorg. Nucl. Chem.* **31** 3783
Reed S J B 1965 *Geochim. Cosmochim. Acta* **29** 513
Reedy R C and Arnold J R 1972 *J. Geophys. Res.* **77** 537
Reuter J H, Epstein S and Taylor H P 1965 *Geochim. Cosmochim. Acta* **29** 481
Reynolds J H 1960 *Phys. Rev. Lett.* **4** 8
Reynolds J H and Turner G 1964 *J. Geophys. Res.* **69** 3281
Ringwood A E 1960a *Geochim. Cosmochim. Acta* **20** 1
—— 1960b *Geochim. Cosmochim. Acta* **20** 155
Romig M F and Lamar D L 1963 *Rand Corp. Mem.* RM-3724 ARPA
Rose G 1825 *Ann. Phys.* **4** 173
—— 1863 *Phys. Abh. Akad. Wiss. Berlin* **23** 30
Rosman K J R and DeLaeter J R 1976 *Nature* **261** 216
Ross J E and Aller L H 1976 *Science* **191** 1223
Rowe M W and Kuroda P K 1965 *J. Geophys. Res.* **70** 709
Russell H N 1929 *Astrophys. J.* **70** 11
Schnetzler C C and Philpotts J A 1969 *Meteorite Research* ed P M Millman (Dordrecht, Holland: Reidel) p 206
Schnetzler C C, Pinson W and Hurley P 1966 *Science* **151** 817
Schramm D N, Tera F and Wasserburg G J 1970 *Earth Planet. Sci. Lett.* **10** 44
Schreibers C von 1820 *Bëytrage zur Geschichte und Kentniss meteorischer Stein- und Metall-massen* (Vienna: Heubner)
Schultz L and Hintenberger H 1967 *Z. Naturf.* **A22** 773
Schultz L, Phinney D and Signer P 1973 *Meteoritics* **8** 435
Schultz L, Signer P, Lorin J C and Pellas P 1972 *Earth Planet. Sci. Lett.* **15** 403
Schultz L, Signer P, Pellas P and Poupeau G 1971 *Earth Planet. Sci. Lett.* **12** 119
Scott E R D 1972 *Geochim. Cosmochim. Acta* **36** 1205
—— 1977 *Geochim. Cosmochim. Acta* **41** 349
Scott E R D and Wasson J T 1975 *Rev. Geophys. Space Phys.* **13** 527
Sears D W 1974 *PhD Thesis* University of Leicester
—— 1975a *Earth Planet. Sci. Lett.* **26** 97
—— 1975b *Meteoritics* **10** 215
—— 1978 *Earth Planet. Sci. Lett.* in press (see *Meteoritics* **12** 362)
Sears D W and Axon H J 1976 *Meteoritics* **11** 97
Sears D W and Mills A A 1973 *Nature Phys. Sci.* **242** 25
—— 1974 *Earth Planet. Sci. Lett.* **22** 391
Shoemaker E M 1963 *The Solar System IV: The Moon, Meteorites and Comets* ed B M Middlehurst and G Kuiper (Chicago: Chicago University Press) p 301

Signer P 1964 *Origin and Evolution of Atmospheres and Oceans* ed P S Brancazio and A G W Cameron (New York: Wiley) p 183

Signer P and Nier A O 1960 *J. Geophys. Res.* **65** 2947

Signer P and Suess H E 1963 *Earth Science and Meteorites* ed J Geiss and E D Goldberg (Amsterdam: North-Holland) p 241

Smales A A, Mapper D and Fouche K F 1967 *Geochim. Cosmochim. Acta* **31** 673

Smith C S 1962 *A History of Metallography* (Chicago: Chicago University Press)

Smith J W and Kaplan I R 1970 *Science* **167** 1367

Sonnett C P, Colburn D S, Schwartz K and Keil K 1970 *Astrophys. Space Sci.* **7** 446

Sorby H C 1864 *Phil. Mag.* **28** 157

—— 1877 *Nature* **15** 495

Stauffer H and Honda H 1962 *J. Geophys. Res.* **67** 3503

Stolper E 1977 *Geochim. Cosmochim. Acta* **41** 587

Studier M H, Hayatsu R and Anders E 1968 *Geochim. Cosmochim. Acta* **32** 151

—— 1972 *Geochim. Cosmochim. Acta* **36** 189

Suess H E, Wänke H and Wlotzka F 1964 *Geochim. Cosmochim. Acta* **28** 595

Tanenbaum A S 1967 *Earth Planet. Sci. Lett.* **2** 33

Tatsumoto M, Knight R J and Allegre C J 1973 *Science* **180** 1279

Tatsumoto M, Knight R J and Doe B R 1971 *Proc. 2nd Lunar Sci. Conf.* **2** 1521 (Cambridge, Mass.: MIT Press)

Tatsumoto M and Unruh D M 1975 *Meteoritics* **10** 500

Taylor H P, Duke H N, Silver L T and Epstein S 1965 *Geochim. Cosmochim. Acta* **29** 489

Taylor H P and Epstein S 1970 *Proc. Apollo 11 Lunar Sci. Conf.* (Oxford: Pergamon) p 1613

Taylor S R and Kaye M 1969 *Geochim. Cosmochim. Acta* **33** 1083

Trivedi B M P and Goel P S 1973 *J. Geophys. Res.* **78** 4853

Tschermak G 1883 *Sitzungsber. Akad. Wiss. Wien, Abt. 1* **88** 347

Turner G 1969 *Meteorite Research* ed P M Millman (Dordrecht, Holland: Reidel) p 407

Urey H C 1952a *The Planets* (New Haven: Yale University Press)

—— 1952b *Proc. Natn. Acad. Sci. USA* **38** 351

—— 1953 *13th Int. Congr. Pure and Applied Chemistry: Plenary Lectures* (Uppsala: Almqvist and Wiksells) p 188

—— 1961 *J. Geophys. Res.* **66** 1988

—— 1962 *The Moon* (*IUA Symp. No. 14*) (New York: Academic Press) p 133

—— 1966 *Mon. Not. R. Astron. Soc.* **133** 199

Urey H C and Craig H 1953 *Geochim. Cosmochim. Acta* **4** 36

Van Schmus W R 1969a *Earth Sci. Rev.* **5** 145

—— 1969b *Meteorite Research* ed P M Millman (Dordrecht, Holland: Reidel) p 480

Van Schmus W R and Hayes J M 1974 *Geochim. Cosmochim. Acta* **38** 47

Van Schmus W R and Wood J A 1967 *Geochim. Cosmochim. Acta* **31** 737

Vdovykin G P 1967 *NASA Tech. Transl.* TTF-582

—— 1970 *Space Sci. Rev.* **10** 483

Von Michaelis H, Ahrens L H and Willis J P 1969 *Earth Planet. Sci. Lett.* **5** 387

Voshage H 1967 *Z. Naturf.* **A22** 477

Wahl W 1952 *Geochim. Cosmochim. Acta* **2** 91

Wai C M and Wasson J T 1977 *Earth Planet. Sci. Lett.* **36** 1

Walter L S 1969 *Meteorite Research* ed P M Millman (Dordrecht, Holland: Reidel) p 191

Wänke H 1965 *Z. Naturf.* **A20** 946

—— 1966 *Fortschr. Chem. Forsch.* **7** 322

Wänke H, Baddenhausen H, Palme H and Spettel B 1974 *Earth Planet. Sci. Lett.* **23** 1

Wasserburg G J, Huneke J C and Burnett D S 1969 *J. Geophys. Res.* **74** 4221

Wasserburg G J, Sanz H G and Bence A E 1968 *Science* **161** 684

Wasson J T 1970 *Icarus* **12** 407

—— 1972 *Rev. Geophys. Space Phys.* **10** 711

—— 1974 *Meteorites* (Berlin: Springer)

—— 1977 *Earth Planet. Sci. Lett.* **36** 21

Wasson J T and Chou C L 1974 *Meteoritics* **9** 69

Wasson J T and Goldstein J I 1968 *Geochim. Cosmochim. Acta* **32** 329

Wasson J T, Schaudy R, Bild R W and Chou C-L 1974 *Geochim. Cosmochim. Acta* **40** 1449

Wasson J T and Wai C M 1970 *Geochim. Cosmochim. Acta* **34** 169

—— 1976 *Nature* **261** 114

Weber H W, Hintenberger H and Begemann F 1971 *Earth Planet. Sci. Lett.* **13** 205

Wetherill G W 1968 *Science* **159** 79

—— 1971 *Physical Studies of Minor Planets* ed T Gehrels *NASA Spec. Publ.* SP 267

—— 1975 *Ann. Rev. Nucl. Sci.* **25** 283

—— 1976 *Geochim. Cosmochim. Acta* **40** 1297

Whipple F 1966 *Science* **153** 54

—— 1971 *Physical Studies of the Minor Planets* ed T Gehrels *NASA Spec. Publ.* SP 267 p 251

Wiik H B 1956 *Geochim. Cosmochim. Acta* **9** 279

—— 1969 *Commun. Phys. Math.* **34** 135

Wilkening L, Lal D and Reid A M 1971 *Earth Planet. Sci. Lett.* **10** 334

Williams J G 1973 *EOS* **54** 233

Willis J and Wasson J T 1977 *Meteoritics* **12** 388

Wolman Y, Haverland J and Millar S L 1972 *Proc. Natn. Acad. Sci.* **69** 809

Wood J A 1962 *Geochim. Cosmochim. Acta* **26** 739

—— 1963 *The Solar System IV: The Moon, Meteorites and Comets* ed B M Middlehurst and G Kuiper (Chicago: Chicago University Press) p 337

Wood J A 1964 *Icarus* **3** 429
—— 1967 *Icarus* **6** 1
Yavnel A A 1958 *Int. Geol. Rev.* **2** 380
Zähringer J 1962 *Z. Naturf.* **A17** 460
—— 1966 *Earth Planet. Sci. Lett.* **1** 379
—— 1968 *Geochim. Cosmochim. Acta* **32** 209
Zimmerman P D and Wetherill G W 1973 *Science* **182** 51
Zittel K A von 1901 *History of Geology and Palaeontology* (London: Ogilvie-Gordon)

Index

Meteorite names are given in italics.

184